Design Methodology for RF CMOS Phase Locked Loops

For a listing of recent titles in the *Artech House Microwave Library*, turn to the back of this book.

Design Methodology for RF CMOS Phase Locked Loops

Carlos Quemada
Guillermo Bistué
Iñigo Adin

ARTECH HOUSE
BOSTON | LONDON
artechhouse.com

Library of Congress Cataloging-in-Publication Data
A catalog record for this book is available from the U.S. Library of Congress.

British Library Cataloguing in Publication Data
A catalogue record for this book is available from the British Library.

ISBN-13: 978-1-59693-383-5

Cover design by Yekaterina Ratner

685 Canton Street
Norwood, MA 02062

10 9 8 7 6 5 4 3 2 1

Contents

Preface

Over the past two decades CMOS technology has become a focus of interest for analog applications. Because of its low power consumption and high integration characteristics, CMOS was initially designed for use in digital applications. However, this situation is rapidly changing due to the continuous development and optimization of new submicron technologies. At present, CMOS technologies are a suitable solution for the integration of low-cost, low-power analog front ends such as WiFi, Bluetooth, Global Positioning System (GPS), ZigBee, or Radio Frequency IDentification (RFID). Another growth factor in this area has been the accessibility of standard processes as virtually all designers have access to technologies up to 100 nm through different multiproject wafers (MPWs).

Within this ever changing environment the methodology to implement circuits becomes crucial. The methodology is the means to guarantee reliable products that meet budgetary and time-to-market constraints, and herein lies the disadvantage to CMOS solutions. Because CMOS was designed to address digital applications; designers are fully aware of the technologies' limitations and understand that they must be innovative and creative in applying these technologies to new opportunities in the market.

This book is designed to act as a practical design guide for CMOS PLL's designers. Phase-locked loops are circuits commonly found in communication front ends, and are especially relevant in OFDM-based systems. Within in its composition there is a wide variety of circuits ranging from digital to RF analog to make this a unique building block. Its diversity allows the student to become familiar with a variety of design techniques and for professionals it remains a challenging task.

We have no desire to duplicate current available information in existing publications. We believe our book documents in clear sequential order practical information from published works. Our sources have included published scientific papers and international conference presentations. This coupled with our own experience as designers and lecturers give the reader a complete detailed account of the pros and cons experienced using the different approach methods. Our goal is to simply bridge what we believe to be, the gap between CMOS technology and PLL theory books.

Chapters 1 to 5 present the different alternatives available for RFIC designers in order to tackle the PLL implementation with CMOS technologies. The structure of these chapters follows the building block architecture of PLL. For each block, the principle of operation is presented, introducing the CMOS realizations available in bibliography with references and discussion of previous works. We have included the phase noise analysis at this juncture of the book as it is one of the most important parameters in the design process. In Chapter 5 we present layout considerations with photographs of our own designs marking the key points of the design.

Chapters 6 to 11 outlines the design case and the characterization setup. These chapters are geared to students and novel designers. Its objective being the illustration of the procedures used to develop a complete PLL, using the methodology, building blocks and models previously presented.

We set out to write a book that would enhance existing available works and act as a valuable learning tool for designers. In a technological climate of change, we as engineers must consistently be innovative in our approach and avail of all opportunities for learning new applications for existing technologies. In bridging the gap between CMOS technology and PLL theory books, we believe the sequence of information and layout of this publication makes it a valuable additional learning tool. In conclusion, we hope you find the book educational, challenging and helpful.

Acknowledgments

We would like to acknowledge the Conserjería de Educación, Universidades e Investigación of the Basque Government for the continuous support to our research group.

We would also like to express our gratitude to all the staff at CEIT and TECNUN (University of Navarra) and our colleagues at University of Sevilla, IUMA Research Centre, and IKERLAN for the fruitful collaboration throughout the last years.

Finally, we would like to mention the people who helped us in the translation and review of the manuscript: Noelene Rippin and Susan Alustiza, and the staff and technical reviewers of Artech House for their support.

1

Approach to CMOS PLL Design

Over the past few years, the communication standards proliferation has been and continues to be motivated by the growing demand of data transmission within the domestic and commercial market. Most commercial solutions offer wireless connections to meet people's increased mobility. To date, the predominant wireless linked networks are working within the 2.4 ISM GHz band. These are based on Bluetooth and WiFi IEEE 802.11b/g standards. However, the necessity of higher data rates and the saturation of the radio spectrum have caused the migration of the newest Broadband Wireless Standards up to higher frequency bands. More specifically, the newer WiFi and the higher range WiMAX networks occupy the licensed 3.5-GHz band and the unlicensed 5 GHz U-NII band. This last frequency allocation extends the bandwidth up to 555 MHz separated in four U-NIIx subbands. The higher amount of users available and the interference-free spectrum used are the main advantages of this newest unlicensed allocation.

The essential requirement to ensure commercial success for the new devices is the achievement of a low-cost price. Its first condition is thereby the fabrication based on silicon technologies such as CMOS. These processes have been traditionally optimized for digital applications, which have given an even higher importance to the design phase. Furthermore, the increased need for mobility within modern communications requires the devices to minimize their power consumption and in turn, this demand increases the restrictions for the high-frequency integrated circuits design. It should also be noted the specifications of these analog blocks are toughened by the newest modulations and multiplexation techniques that are based on the orthogonal frequency division multiplexing (OFDM) method.

In many of these last generation communication systems, the phase locked loop (PLL) has become the standard solution for the implementation of frequency synthesizers. The PLL itself can be considered as a unique system and its design is one of the most challenging requiring focused efforts. This assertion is based on the knowledge that PLL is a combination of several building blocks with different characteristics covering the entire spectrum of circuit design: analog RF, mixed-signal, and digital. Obviously the strategies and techniques required to implement each circuit varies and each individual building block requires a specific approach. This approach has to be driven by an appropriate design flow that ensures that the final design meets the established requirements.

The objective of this chapter is to review the basic concepts related to CMOS technology and the inherent problems designers face implementing PLLs using it. Sections 1.1 and 1.2 briefly present MOS basic models and effects needed to predict transistor behavior. Sections 1.3 and 1.4 are key to understanding the core issues on CMOS PLL design. Here we present the practical problems that an IC designer faces when trying to work with real-case circuits such as technologies usually optimized for digital applications, lack of accurate models for passive components (inductors and varactors), or no access to the parameters of the technology. We take a worst-case scenario design viewpoint and offer methodology to overcome the problems. Our final point in this introductory chapter is the PLL design flow proposed in our work. This design flow is used in Chapters 6 through 10 to design a complete PLL using the information provided in previous chapters.

1.1 MOS Transistor Basics

The following sections present a review of basic theory about transistor NMOS. This study is designed to alert the reader to the most relevant concepts that should be taken into account when dealing with this type of transistor.

The symbols adopted to depict the transistors are shown in Figure 1.1.

The next section presents the basic structure of these components. Note that the discussion is centered on enhancement type transistors, because these are the most common devices included in RF analog integrated circuits.

1.1.1 Enhancement Type MOSFET Structure

Figure 1.2 shows the basic diagram of the N-channel enhancement type MOSFET. This device, in its simpler approach, consists of two n-type regions (drain and source) separated by a P-type channel. The gate electrode, usually a polysilicon layer, is isolated from the substrate by means of a thin oxide layer.

The P-channel enhancement MOSFETS(PMOS) transistors are composed by a complementary structure. However, the practical implementation with

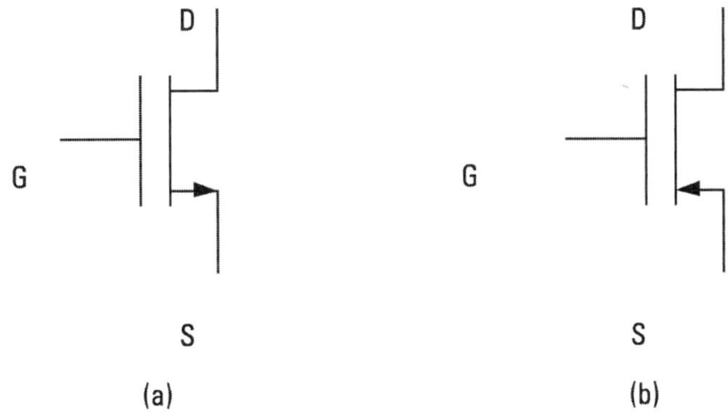

Figure 1.1 MOS transistor symbols. (a) N-channel enhancement NMOS and (b) P-channel enhancement NMOS.

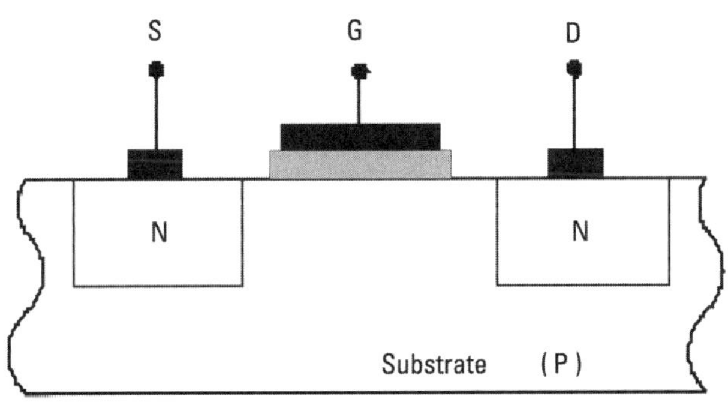

Figure 1.2 Cross-section of the N-channel enhancement-type MOSFET.

CMOS technologies results in some important differences, shown in Figure 1.3.

As it can be seen in the previous figure, in integrated technologies both P and N transistors share the same substrate. Thus PMOS device must be embedded in an N-well. During normal operation, this well is usually connected to an appropriate potential such as source and drain junction diodes and remain reverse-biased. However, this terminal adds extra flexibility for the designers, which is especially useful in analog design.

Another important difference between NMOS and PMOS transistors comes from the fact that the characteristics of the respective channels are usually very different. The channel mobility of the holes of the P-channel is approximately three times less than that of the electrons in N-channel. Although this figure may change for the different fabrication processes, NMOS are generally favored. Therefore, they are the component of choice for many designs. For

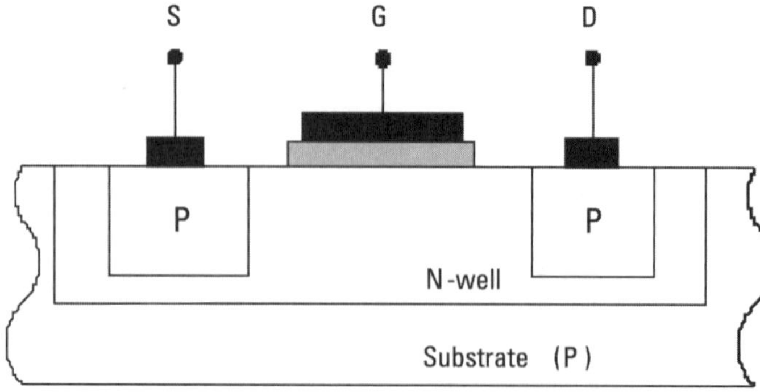

Figure 1.3 Cross-section of the P-channel enhancement-type MOSFET.

this reason, the next subsection explains the operating principles of NMOS devices.

1.1.2 Operating Principles of the N-Channel MOSFET Transistor

With the configuration shown in Figure 1.2, no significant current is expected to flow between terminals D and S, due to the fact that a reverse-biased PN junction is always found in any possible current path. However, we can change this situation by applying an appropriate gate voltage, as shown in Figure 1.4.

When a positive voltage is applied to the gate the free holes under the gate electrode (the channel region) are repelled, whereas electrons from source and drain regions are attracted. When a significant number of charge carriers have been redistributed, the region under the gate contact is inverted to an N-channel, thus allowing current flow between gate and drain contacts. The

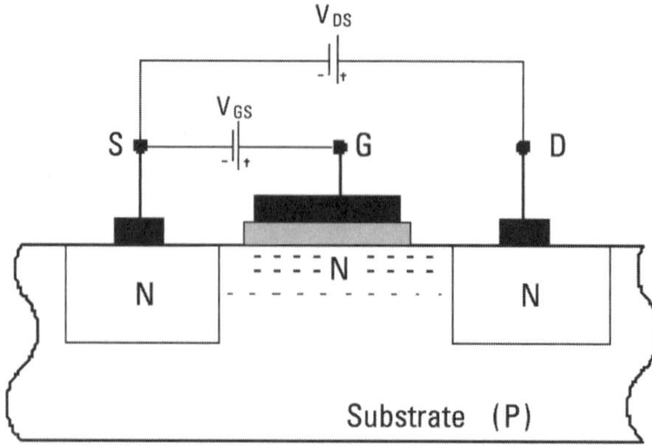

Figure 1.4 N-channel induced by positive voltage applied to the gate.

minimum value of V_{GS} for which this phenomenon occurs is called the threshold voltage (V_{TH}). Figure 1.5 sketches the I_D versus V_{GS} curve, showing this effect.

Following the graph shown in Figure 1.5, the classic circuit design theory divides the operation of the transistor into two zones: the subthreshold zone, where the expected current is negligible, also called the cutoff region, and the "on" state, which is suitable for linear circuit design. However, driven by the growing interest in low power consumption circuits over recent years, the weak inversion zone has gained the interest of circuit designers, due to its good gain efficiency in terms of consumed current and its high linearity.

Figure 1.6 shows an ideal approximation to I_D versus V_{DS} behavior. For V_{DS} values below V_{GS}–V_{TH}, the current I_D follows a parabolic law. When the V_{GD} voltage is high enough to pinch off the channel under the vicinity of drain contact, the value of I_D tends to stabilize, ideally showing no influence of further V_{DS} increases.

The simplest model of this behavior follows a continuous square law for nonlinear and linear triode:

$$I_D = \mu_n C_{ox} \frac{W}{L} \left[(V_{GS} - V_{TH}) V_{DS} - \frac{1}{2} V_{DS}^2 \right] \tag{1.1}$$

The expression for the saturation region can be written as follows:

$$I_D = \frac{1}{2} \mu_n C_{ox} \frac{W}{L} (V_{GS} - V_{TH})^2 \tag{1.2}$$

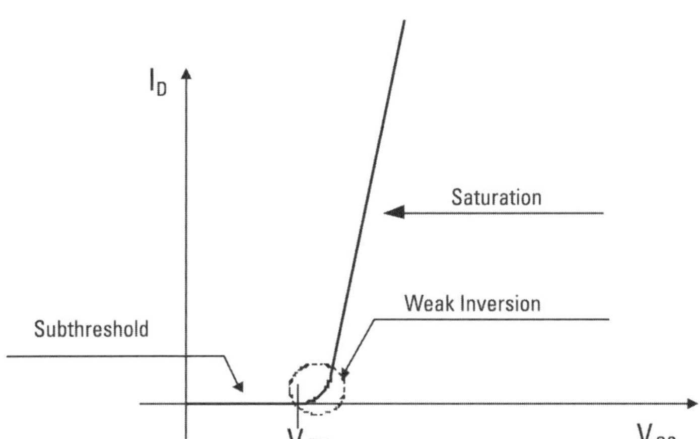

Figure 1.5 The I_D–V_{GS} characteristic of the N-channel enhancement-type MOSFET.

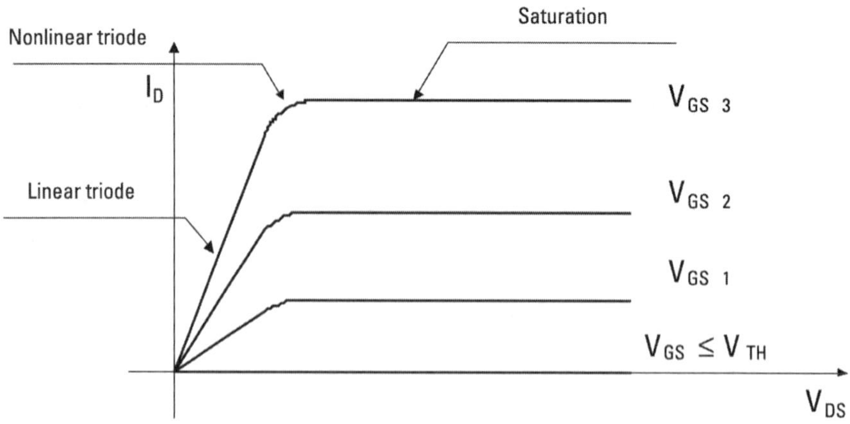

Figure 1.6 The ideal I_D–V_{GS} characteristic of the N-channel enhancement-type MOSFET.

In these expressions:

- μ_n = Electron mobility;
- C_{ox} = Capacity per unit gate area;
- W = Channel width;
- L = Channel length;
- V_{TH} = Threshold voltage.

From the circuit designer point of view, especially when using standard technologies, it is very interesting to note the dependence of the magnitudes with transistor size (W and L), since the rest of parameters are usually fixed. The previous expressions also allow for the calculation of the transconductance, another key magnitude, defined as the ratio between I_D and V_{GS} changes. In the particular case of saturation region, the resulting expression is:

$$g_m = \mu_n C_{ox} \frac{W}{L} (V_{GS} - V_{TH}) \tag{1.3}$$

For subthreshold operation, (1.4) gives a better approximation:

$$g_m = \frac{I_D}{\zeta V_T} \tag{1.4}$$

where ζ is a nonideality factor and V_T is the thermal voltage.

Obviously, the real behavior of NMOS transistors is more complex. However, this simple approach is very useful in order to obtain a basic idea of the characteristics of the circuit under study.

1.2 Nonideal and Second-Order Effects in Submicron Technologies

The physical implementation of NMOS transistors with CMOS technologies leads to several deviations from the ideal behavior shown in Section 1.1. The following outlines the most relevant effects that should be observed during the first stages of design.

Obviously, the inclusion of these effects to the behavioral expressions of the transistor results in the increase of the complexity of the calculations, making unavoidable the use of computers and CAD tools. As many authors have pointed out before, circuit simulation is an art unto itself. However, it is worth mentioning that the approach to circuit design must be always progressive, from the simplest "hand" models to accurate advanced circuit simulation. The first step contributes to develop general ideas about the circuit under study, while the latter refines the solution. Both stages are necessary and complementary.

Most of the simulators currently available are based on SPICE Level 3 models (SPICE-3), which offer a good approximation for discrete circuits and transistors with gate lengths larger than 1 μm. Although SPICE-3 includes some capacitance terms, high frequency simulations of submicron devices requires more advanced models such as BSIM3, BSIM4, or MODEL9. Perhaps BSIM3 is the most popular, as it has been adopted by many foundries in their design kits. Despite not being developed specifically for HF simulations, if the designer knows its limitations and provides extra components to complete the model when necessary, it is possible to obtain accurate results (at least for transistors with gate length higher than 100 nm).

Finally, we would like to remark that the objective of this chapter is to summarize the basic concepts involved in design.

1.2.1 Channel-Length Modulation

This phenomenon is especially important in short channel devices, as is the case in integrated CMOS technologies. It is caused by the variation of the effective channel length in saturation mode, due to the pinch-off of the inversion layer under the drain contact, as depicted in Figure 1.7.

The impact on the transistor behavior is shown in Figure 1.8. Note that the reduction of the effective channel length leads to an increase of the drain current over the expected ideal value. This effect is usually quantified through

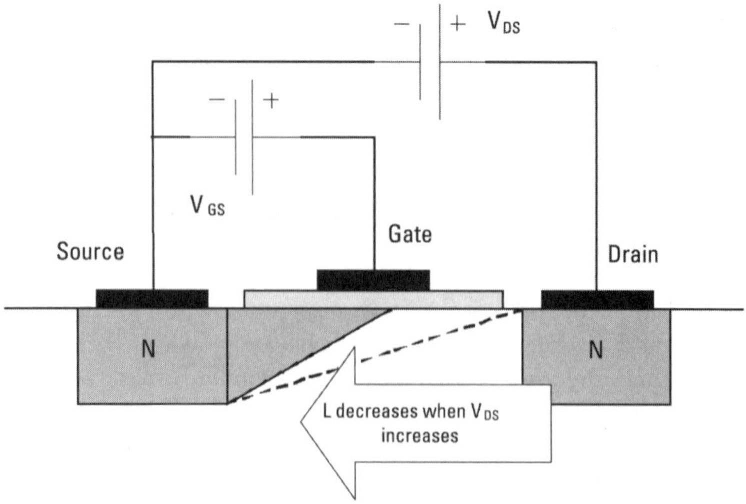

Figure 1.7 Channel length modulation in NMOS transistors.

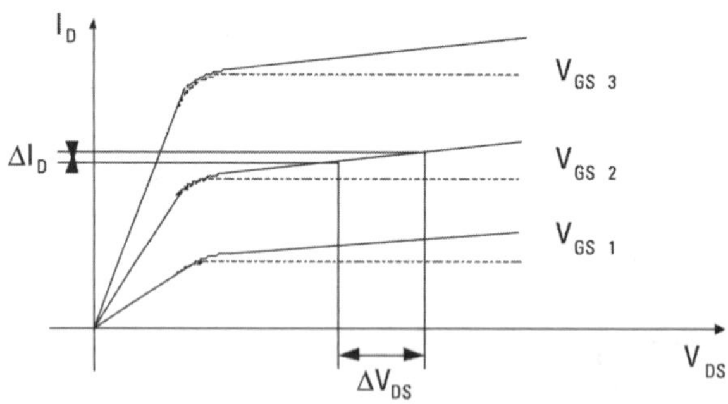

Figure 1.8 Finite output resistance of NMOS transistors.

the parameter called output resistance, which is defined as the change of I_D divided by the change of V_{DS}. This parameter is described in SPICE-3 using the LAMBDA parameter.

The impact of this effect on the output parameters of a circuit can be minimized using transistors with the highest length possible. This causes a problem for RF applications, as larger transistors exhibit worse high-frequency figures of merit. Therefore, the selection of the appropriate transistor size is one of the most important decisions to be made. Usually, the foundries provide different sets of devices that must be analyzed carefully before the actual design starts. In these cases it is very useful to identify the different functions of the transistors of our circuit (e.g., current source, switches, RF input stage or high-power output stage) and match this list with the real transistors available in the design kit.

1.2.2 Parasitic Capacitances

Due to the physical implementation of NMOS transistors using semiconductors, the parasitic capacitances are one of the most important limiting effects that must be considered. To get an idea of the complexity of this problem, just a basic analysis of the device structure reveals the existence of at least six parasitic components, as shown in Figure 1.9.

There are three main categories of capacitive effects:

1. *Junction capacitances.* During the normal operation of the transistor, the source and drain regions are reverse-biased with respect to the substrate. Therefore the corresponding capacitances C_{jSB} and C_{jDB} are found there. These depend on the layout (width and length) but they are also bias-dependent.
2. *Overlap capacitances.* This effect is due to the unavoidable lateral diffusion of source and drain regions (C_{ovGS} and C_{ovGD}). It is determined by the fabrication process, and from the point of view of design, it is proportional to the transistor's width.
3. *Channel capacitances.* The gate contact and the inversion layer form a bundle of conductors, oxide, and semiconductors, leading to several additional capacitances (C_{ChG} and C_{ChB}). As a rule of thumb, we can consider these capacitances proportional to the product W L.

The contribution of these different effects to the overall performance of the device depends on different factors, including the technology characteristics,

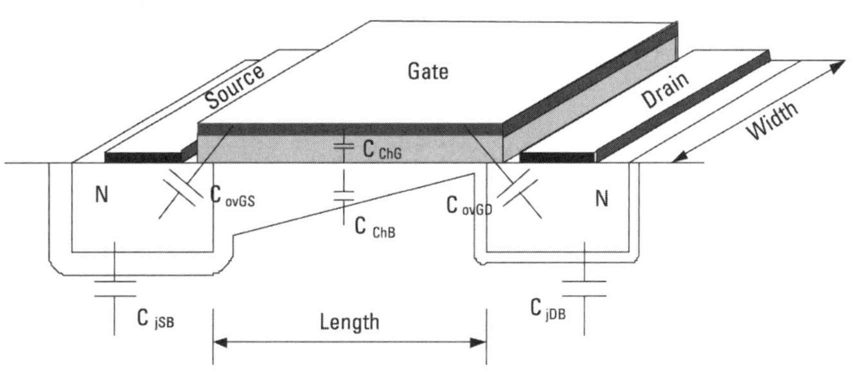

Figure 1.9 Main parasitic capacitances of NMOS structure.

the transistor's size, or even the bias conditions. In [1], an extensive study can be found, with a useful table of the contribution of each parasitic capacitance as a function of bias zone.

Although it is obviously very important to understand the origin of each of these components, it must also be noted that often the circuit designer does not have all the input data needed to calculate the actual values of them. In these cases, using simpler models is highly recommended. The two capacitances of the high-frequency model depicted in Figure 1.10 are one of the two most commonly used. This simplified model has been obtained neglecting body effect and thus the junction capacitances.

The values of C_{GS} and C_{GD} can be estimated with the expressions shown in Table 1.1. Those values should be augmented with the overlap capacitance, although this term is often negligible.

The graphical representation of these capacitances as a function of the different regions of operation can be found in Figure 1.11.

1.2.3 Gate Resistance

In low-frequency models, the gate contact is often considered to be an open circuit, since the oxide layer effectively blocks the current. This situation changes

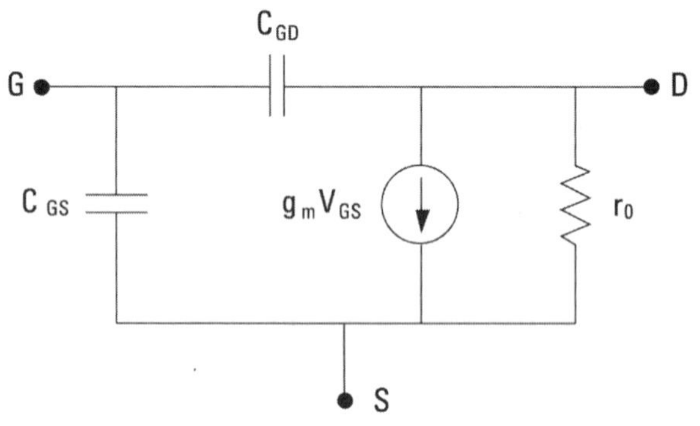

Figure 1.10 Simplified high-frequency model of NMOS transistor.

Table 1.1
Estimation of Capacitances of the High-Frequency Model

	Cutoff	Saturation	Triode
C_{GS}	0	$\frac{2}{3}WLC_{ox}$	$\frac{1}{2}WLC_{ox}$
C_{GD}	0	0	$\frac{1}{2}WLC_{ox}$

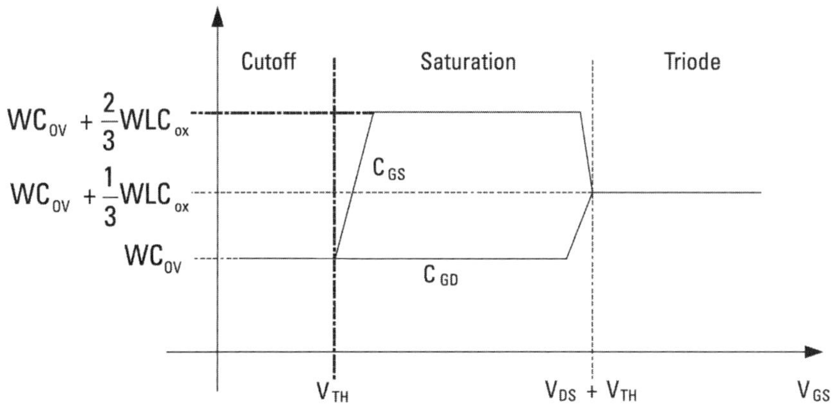

Figure 1.11 Variation of HF model capacitances versus V_{GS}.

dramatically for high-frequency operation. The effect of parasitic capacitances allows current to flow through the gate contact, therefore the series resistance of the gate material must also be modeled.

While some of the last generation transistor models include accurate gate resistance parameters, it must be noted that early BSIM3v3 versions did not take into account this effect. Therefore, the designer must be aware of this problem, and in the case of using an early model this effect should be introduced as an extra component. Regardless of the availability of this effect in the design kit, it is always interesting to have an idea of the value of this resistance, and this task can be accomplished by looking at several published studies. For example, in the work presented by Kolding et al. [2], the gate contact is represented by a distributed transmission line, leading to a set of simple expressions to calculate gate resistance (Table 1.2).

In this table, R_S is the sheet resistance of the polysilicon gate (typically set by the process and therefore, the designer usually doesn't have control over this value), W is the transistor's width and N is the number of fingers of the transistor. As it is apparent from these expressions, the impact of gate resistance can be minimized with proper layout techniques such as terminal folding. However, the designer must ensure the compatibility of this technique with the particular design rules of the technology used, and it is also crucial to take into account the effect of the folded contacts on the overall circuit implementation.

1.2.4 Body Effect

The body effect arises when source and bulk terminals are not connected to the same voltage This situation is found in circuits with stacked transistors, where the bulk is connected to power rails to prevent internal currents.

Table 1.2
Gate Resistance of NMOS Transistor

		Gate Resistance
Single gate	Single contact	$\frac{R_S W}{3}$
	Double contact	$\frac{R_S W}{12}$
Interdigit gate (N fingers)	Single contact	$\frac{R_S W}{3N}$
	Double contact	$\frac{R_S W}{3N}$

From a practical point of view, the most apparent consequence of body effect is the variation of the effective threshold voltage. This variation is bias-dependent, and obviously it can be modulated tracking the variations of source voltage, hence it can be source of degradation of RF performance.

Although this effect is well known and the impact on the circuit can be easily estimated, the designer must take into account the combination of this effect with the substrate coupling with other components, which are not often correctly modeled in the design kits. In these cases, the only resource available to improve the situation is the use of appropriate layout design techniques.

Some practical expressions for the complete understanding of this effect can be found in the work of Caverly [3]. Here the cross influence of bulk, substrate, and threshold voltage is discussed, and a useful circuit model is proposed.

1.3 Impact on PLL Performance

The real effects of the generic CMOS technologies presented above induce nonidealities in the PLL performance. What is a PLL? And what are the nonideal effects that distinguish these elements?

The concept behind phase-locked loops (PLLs) is explained in Chapter 2 where the different structures are presented. But, in essence, a PLL is a block that generates a controlled frequency tone for the downconversion and/or upconversion of an RF signal. This tone has many characteristics that depend on the application and the standard considered (i.e., a tuned or fixed frequency or the output power).

Its block diagram varies from one structure to another, but the basic scheme includes an oscillator and a fixed and stable reference, normally crystal quartz. These two frequency tones are compared and the signal error generated is then sent to the oscillator tuning pin that controls the output signal of the PLL. In the majority of the PLLs the comparison requires a high-frequency divider for the VCO output since the common and affordable quartzes work in the megahertz band, while the RF VCOs may reach several gigahertz. As it will be extended in Chapter 4, the complete divider is composed of several stages with the first ones at the highest speeds in comparison to the lower frequency dividers where more standard digital techniques can be employed. These high-frequency dividers are the main challenge from the design point of view.

As explained in Section 1.2, the CMOS transistors suffer from parasitic effects that provoke nonidealities in any active circuits. Evidently these active parts are the core of any circuit, and consequently this causes effects on their behavior. Furthermore similar real concepts could be extrapolated for the passives obtained by CMOS technologies, which are also main actors in the circuits' composition.

In the next sections we will outline the impact of these nonidealities in phase noise and general PLL behavior.

1.3.1 Phase Noise

This is the main disturbance of the analog circuits of a frequency and phase-oriented block. This concept is largely extended in the next chapter and afterwards specifically applied for each block of the PLL, but simplifying, this can be defined as the random deviation of a frequency tone that is spread around the center frequency as Figure 1.12 depicted. The theoretical origins and formulations are there detailed for a good understanding.

In RF applications the ideal tone of a PLL is usually characterized by an impulse, whereas the real one is spread in the frequency domain as shown in the right side of Figure 1.12. This affects the sided frequencies, and this offset is important as the phase noise is defined by means of the amplitude difference

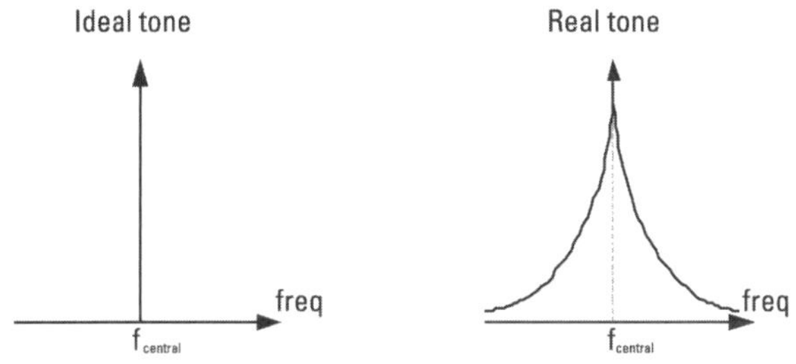

Figure 1.12 Phase noise effect on an ideal frequency tone.

depending on the offset to the center frequency. The unit of the amplitude is normally dBc, which provides a global unit of dBc/Hz at *x* Hz, where *x* is the frequency offset.

It is clear from this Figure 1.12 that this nonideality represents a serious threat in RF communications where exact frequency translations are required. This would be the case with the ideal tone, but the real one's skirt may corrupt the up- or downconversion in the transmitter or receiver. For example a wanted signal and an interferer situated at 1 kHz from the center of the synthesizer tone mixed with this tone which phase noise is –20 dBc/Hz at 1 kHz would produce two output tones: the desired output and an unwanted one 20 dB lower at 1 kHz. Wider explanation about the effects of the phase noise on generic receiver and transmitter is widely found in the extended design bibliography supplied in Section 1.6.

Another concept bound to the phase noise is the jitter, which in its basic meaning is the time domain counterpart of the phase noise. It then affects the blocks where the inputs, outputs and operations are mainly referred to this variable (i.e., the low frequency dividers).

A priori the main actor in the PLL phase noise definition seems to be the oscillator which creates the output tone, but as it is presented in this book, other contributors have to be taken into account for a complete overview of this effect. As it has been yet introduced, in the following chapters wider explanations about this issue is discussed.

1.3.2 Parasitic Capacitances and PLL Behavior

The other main non ideality concerning the CMOS PLLs are the parasitic capacitances of the technology. These not only affect the frequency output of the PLL but also intermediate nodes which may provoke delays and a slowing down in the tuning of the wanted frequency.

The first and principal effect is due to the fact that most of the PLLs designed for RF communications have LC-tank oscillators in their loop. These oscillators present quite large impedance in their load at the resonant frequency of the LC-tank and any extra capacitance added by the transistors or by the final layout of the circuit affects the output frequency. This is one of the clues that have to be considered from the very beginning stages of the design in order to minimize the redesign time. For this, an accurate modeling of the transistors and the passives elements need to be simulated and some basic layout techniques are required, as they are presented along this book.

The second parasitic capacitance effect is not as evident as it affects intermediate blocks of the PLL loop. These blocks are the high frequency circuits, primarily the frequency divider. A deficient design and implementation of the required exact division of the PLL output causes an inadequate control voltage at the tuning pin of the oscillator.

As it is going to be presented in the corresponding chapters, the frequency dividers designed for integrated RF circuits are composed of series switching latches in row. The connections in and among them are crucial in their frequency behavior. The nodes have their own impedance, but the addition of the parasitic capacitances and resistances due to the layout of the design interferes in the switching of these dividers and corrupts the division ratio.

Nevertheless another disturbance is caused by the parasitic capacitance in the comparison blocks. These, as explained in further chapters, have in most of the cases, decision circuits which highly depend on many information data coming from several parts of the system. The ideal behavior of these different signals is to share the time base, but the intermediate capacitances apart from the channels' nonidealities hinder the synchronization.

The dynamics of the PLL loop are taken into account all along this book and specific techniques for the divider and the comparison blocks are presented in the corresponding chapters.

1.4 State of the Art and Challenges in CMOS PLL Design

As briefly outlined in the previous section, the complete PLLs have two main characteristics: The output frequency of oscillation and the phase noise of the output tone.

Nevertheless, other parameters are needed in the definition of a PLL. Table 1.3 presents a summary of the main relevant synthesizers found in the references for the stringent 5-GHz region.

The second column is the CMOS technology used for these designs. The most recent papers present CMOS 0.13 μm designs but concerning the frequency band of interest here selected (around 5 GHz, presented in the second column), from CMOS 0.25 μm, the main objectives can be reached. A more modern technology would improve the higher frequency reached, but there are also other details to take into account, as seen later. Technologies below 100 nm suffer from severe flicker noise, due to the hot electron effect.

The third and fourth columns depict the evolution of the phase noise of these PLLs. All the phase noise in dBc/Hz has been transposed to 1 MHz of distance to the center of the tone in order to allow a correct comparison. This is the key parameter of all PLL. As outlined earlier, the more negative the purer the tone resulting in the best mix with this synthesizer. For the frequencies here selected in Table 1.3, the standards in use have a requirement close to −110 dBc/Hz at 1 MHz, which is a stringent data, but achievable for the CMOS technologies.

Table 1.3
CMOS Integrated Synthesizers Working Around the 5-GHz Band

Reference	Tecn. CMOS	Freq. (GHz)	Phase Noise at 1 MHz (dBc/Hz)	∫P. N. (dBc)	*Spurs* (dBc)	V_{dd} (V)	Pow. (mW)
[4]	0.13 μm	5.5	–116.7	–36	—	1.2/2.5	36
[5]	0.18 μm	5.5	–115.7	—	–80 at 11 MHz	1	27.5
[6]	0.18 μm	5.2	–116.2	–34	—	1.8	—
[7]	0.18 μm	3.5	–120.7	–31.7	–66 at 13 MHz	1.8	—
[8]	0.18 μm	5.2	–114.2	—	—	1.8	—
[9]	0.25 μm	4.3	—	—	–58 at 4 MHz	2.5	117.5
[10]	0.18 μm	5.2	–119.2	–32	–65 at 10 MHz	1.8	—
[11]	0.25 μm	4	–114	–33	—	2.5	180
[12]	0.25 μm	5.4	116.0	—	69 at 43 MHz	1.5	23
[13]	0.25 μm	4.9	–104.7	—	–54 at 22 MHz	1.5/2	25

Day by day, more complex modulations are required to perform higher output data rates and this causes the strengthening of the phase noise specification. Fortunately, lower frequency bands are being selected for the allocation of current applications and consequently even better phase noises can be achieves with these CMOS technologies.

The fifth column is dedicated to the spurious tones founded at the center of the adjacent channel. It depends mainly on the standard considered in each case, but the improvement of the technology provides lower spurs. This is due to the distances and the parasitics performed by even the tiniest CMOS technologies allow adjustments to meet designer's goals.

The two last parameters are strongly related, and its optimization is one of the big challenges of the modern RF design. The power consumption is the supply voltage applied to the circuit multiplied by the current consumed by the system. The lowering of the power consumption starts with the use of a lower supply voltage, but this may lead to more stringent specifications. Thereby, a cautious and conscious design flow has to be taken into consideration to improve this data.

1.5 PLL Design Flow

A PLL is a complex system and differs in the design approach from other circuits. For this reason, it is important to observe the different design levels and to ensure that each step forward is firmly based on correct parameters. In the next paragraphs we outline a design flow that could be used for PLL design:

The first step is the determination of the initial requirements of the synthesizer. Despite the fact that the list of requirements can be extensive, it is recommended that you begin the design paying special attention to the main parameters which are:

- Output frequency;
- Reference crystal frequency and its accuracy;
- Specification of phase noise inside and outside the loop bandwidth;
- Maximum level of spurious emissions;
- Lock time of the PLL.

Some of these specifications can be extracted directly from the standard considered for the application, and others need to be deduced using complementary documents.

The following step is to choose the architecture of the frequency synthesizer. Out of the diverse architectures the advantages and disadvantages of each one needs to be evaluated, taking into account the limitations of the fabrication technology that is going to be used for its implementation.

Following this, the specifications of each one of the blocks of the synthesizer must be determined. For this process an iterative methodology should be used in which the input data comes from:

- The designer's previous experience;
- Scientific publications;
- Datasheets.

In order to carry out the calculations, it is essential to use a system simulation program. There are excellent commercial software packages suitable to carry out this function, but it is necessary to point out that it is also possible to create work models in general programs such as MATLAB. In the case presented in this book, a specific tool for the design of the PLL has been developed in-house. The advantage in this approach is that the designer has total control over the number of variables that are introduced in the process like in the models used. Moreover, the reduction of costs is also evident.

The iterative process ends when the specifications selected for the blocks give rise to a system that meets the parameters set out in point 1 of the process of design.

Once the initial simulation of the frequency synthesizer provides results that meet the starting requirements, you can begin with the design of each block separately by means of an electronic circuit's simulation tool, trying to meet the specifications of each block determined in step 3. It is at this point that the inevitable discussion between circuit designers and system managers take place resulting in a trade-off between the diverse circuits for the global optimization of the system.

With the final specifications obtained in step 4, the frequency synthesizer is simulated again with the system software, checking once again that the simulated results meet the starting requirements. In a negative case, it is necessary to return to step 4 and redesign the necessary blocks until meeting the requirements set out in step 1 are accomplished. Figure 1.13 demonstrates the design process described earlier.

1.6 Basic Design Bibliography

The keystone of any complete block design is the circuit design of the different elements. For this purpose, many publications have stated the basis from which the designers can adapt the circuits to their specific goals. The following list is only a representation of the main books on this field:

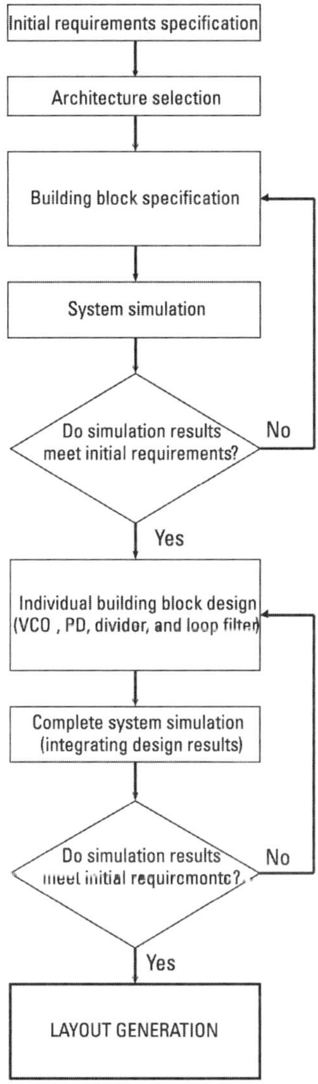

Figure 1.13 Design flow of an integrated frequency synthesizer.

Allen, P. E., and D. R. Holdberg, *CMOS Analog Circuit Design*, New York: Oxford University Press, 1987.

Baker, R. J., H. W. Li, and D. E. Boyce, *Circuit Design, Layout, and Simulation*, New York: IEEE Press, 1998.

Caverly, R., *CMOS RFIC Design Principles*, Norwood, MA: Artech House, 2007.

Craninckx, J., and M. Steyaert, *Wireless CMOS Frequency Synthesizer Design*, Norwell, MA: Kluwer Academic Publishers, 1998.

Gray, P. R., and R. G. Meyer, *Analysis and Design of Analog Integrated Circuits,* New York: Wiley, 1993.

Hastings, A., *The Art of Analog Layout*, Upper Saddle River, NJ: Prentice-Hall, 2001.

Lee, T. H., *The Design of CMOS Radio Frequency Integrated Circuits*, New York: Cambridge University Press, 1998.

Razavi, B., *Design of Analog CMOS Integrated Circuits*, New York: McGraw-Hill, 2001.

Razavi, B., *RF Microelectronics*, Upper Saddle River, NJ: Prentice-Hall, 1998.

Rohde, U. L., and D. P. Newkirk, *RF/Microwave Circuit Design for Wireless Applications*, New York: Wiley-Interscience, 2000.

Sedra, A. S., and K. C. Smith, *Microelectronic Circuits*, 4th ed., New York: Oxford University Press, 1998.

References

[1] Lee, T. H., *The Design of CMOS Radio Frequency Integrated Circuits*, New York: Cambridge University Press, 1998.

[2] Kolding, T. E., *Calculation of MOSFET Gate Impedance*, Technical Report R98-1009, RISC Group, Aalborg University, ISSN 0908-1224, August 1998.

[3] Caverly, R., *CMOS RFIC Design Principles*, Norwood, MA: Artech House, 2007.

[4] Valla, M., et al., "A 72-mW CMOS 802.11a Direct Conversion Front-End with 3.5-dB NF and 200-KHz 1/f Noise Corner," *IEEE Journal of Solid-State Circuits*, Vol. 40, No. 4, April 2005, pp. 970–977.

[5] Levantino, S., et al., "Phase Noise in Digital Frequency Dividers," *IEEE Journal of Solid-State Circuits*, Vol. 39, No. 5, May 2004, pp. 775–783.

[6] Ahola, R., et al., "A Single-Chip CMOS Transceiver for 802.11a/b/g Wireless LANs," *IEEE Journal of Solid-State Circuits*, Vol. 39, No. 12, December 2004, pp. 2250–2258.

[7] Zhang, P., et al., "A 5-GHz Direct-Conversion CMOS Transceiver," *IEEE Journal of Solid-State Circuits*, Vol. 38, No. 12, December 2003, pp. 2232–2237.

[8] Behzad, A. R., et al., "A 5-GHz Direct-Conversion CMOS Transceiver Utilizing Automatic Frequency Control for the IEEE 802.11a Wireless LAN Standard," *IEEE Journal of Solid-State Circuits*, Vol. 38, No. 12, December 2003, pp. 2209–2220.

[9] Herzel, F., G. Fischer, and H. Gustat, "An Integrated CMOS RF Synthesizer for 802.11a Wireless LAN," *IEEE Journal of Solid-State Circuits*, Vol. 38, No. 10, October 2003, pp. 1767–1770.

[10] Vassiliou, I., et al., "A Single-Chip Digitally Calibrated 5.15-5.825-GHz 0.18-μm CMOS Transceiver for 802.11a Wireless LAN," *IEEE Journal of Solid-State Circuits*, Vol. 38, No. 12, December 2003, pp. 2221–2229.

[11] Zargari, M., et al., "A 5-GHz CMOS Transceiver for IEEE 802.11a Wireless LAN Systems," *IEEE Journal of Solid-State Circuits*, Vol. 37, No. 12, December 2002, pp. 1688–1694.

[12] Hung, C.-M., and K. O. Kenneth, "A Packaged 1.1GHz CMOS VCO with Phase Noise of –126 dBc/Hz at a 600KHz Offset," *IEEE Journal of Solid-State Circuits*, Vol. 35, No. 1, January 2000, pp. 100–103.

[13] Rategh, H., H. Samavati, and T. Lee, "A CMOS Frequency Synthesizer with an Injection-Locked Frequency Divider for a 5-GHz Wireless LAN Receiver," *IEEE Journal of Solid-State Circuits*, Vol. 35, No. 5, May 2000, pp. 780–787.

2

PLL Fundamentals

This chapter presents a review of PLL fundamentals. Section 2.1 is dedicated to presenting a building block diagram and the basic formulas of the most common PLL architectures (integer-N and fractional). Section 2.2 discusses the three main figures of merit selected as key points for PLL design: phase noise, spurious signals level, and lock time. Regarding the first one in Section 2.2.1, we present the formulas needed to calculate the PLL behavior taking into account the contribution of each building block. The general formula is shown here, while the origin of individual phase noise of each circuit (VCO, frequency divider, and so forth) is explained later in its corresponding chapter. Section 2.2.2 describes and compares two models available in the references for spurs level calculation in CMOS synthesizers. Finally, in Section 2.2.3 an expression for lock-time calculation is deduced. In each case, the impact of these figures of merit on PLL behavior is also explained, referencing in particular the effect on modern OFDM based communications systems.

2.1 Frequency Synthesizer

In Figure 2.1, the block diagram of a phase-locked loop, also called an indirect frequency synthesizer, can be seen. The working principle of this synthesizer consists of continuously correcting the frequency and/or phase difference that exists between the two periodic input signals to the loop (i.e., F_{xtal} and F_{div}) When there is no difference, it is said that the loop is locked, and the output frequency of the synthesizer (F_{out}) is a multiplication of the reference frequency F_{xtal} by a variable number that is the quotient between the values N and R.

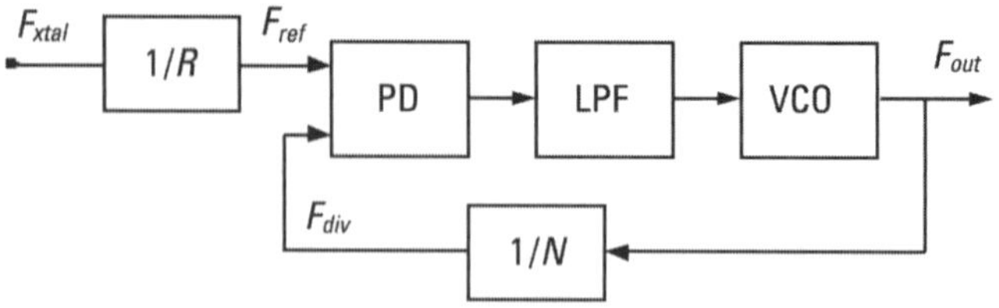

Figure 2.1 Phase-locked loop.

The frequency of the quartz crystal (F_{xtal}) is represented in this figure followed by a divider by R for the sake of the general case. At the same time, the voltage controlled oscillator's (VCO) input is divided by N, in the frequency divider, in order to be able to compare it with the reference signal in the phase detector.

In the phase detector (PD), the divided frequency (F_{div}) is compared with the reference frequency (F_{ref}) generating a proportional signal to the phase difference that exists between the two input signals. Generally, in CMOS PLLs together with the PD, a charge pump (CP) is usually integrated, which is responsible for transforming the PD output signal into a source of current pulse train, dependent on the output. Further discussion on these blocks can be found in Chapters 4 and 5.

The CP output signal is filtered using a lowpass filter (LPF), which converts the current pulse train into a continuous voltage to control the VCO. Indeed, it acts as a transimpedance network to perform the current to voltage conversion as well as the filtering. The frequency response of this LPF impacts the overall PLL performance and its stability. Furthermore, this filter attenuates the undesired spurious emissions and determines the band width of the loop, a parameter that influences the total noise of the PLL.

As a rule of thumb, even in an integrated circuits design context, the elements that composed this LPF can overpass the technological limits given by the foundries and are made of discrete elements. This is a common issue for the researchers in charge of the modern PLLs design. An example of this matter is presented in the chapters dedicated to practical concerns in this book.

When the loop is locked, the two inputs to the phase detector maintain a relation of constant phases and as a consequence the same frequency. The synthesized frequency can be obtained using the following formula:

$$F_{out} = \frac{N}{R} \cdot F_{xtal} \tag{2.1}$$

In this way, possible variations in the input phase signal from the VCO will be transmitted at the input and corrected by means of a phase detector and low-pass filter.

Depending on the division value N, this type of synthesizer is usually classified as an integer synthesizer or a fractional synthesizer. The following briefly explains the operation of each one. The derivation of the transfer function has been omitted in order to simplify the text, but they can be found in some of the references listed at the end of the chapter [1, 2].

2.1.1 Integer-N Architecture

The architecture of this synthesizer is presented in Figure 2.2, where the input frequency divider R has been omitted for simplicity. This type of synthesizer generates a frequency F_{out} that can be calculated according to (2.2), where N varies in unit stages from N_L to N_H.

$$F_{out} = N \cdot F_{ref} \tag{2.2}$$

The key point in this architecture is that the reference frequency (F_{ref}) must coincide with the spacing between channels in order to allow consistent channel selection. In this way, when N takes the value N_L the inferior channel is selected. Later, by means of the variation of N, the rest of the channels are selected successively until the superior channel is reached and which is tuned when N reaches the value N_H. Therefore, the frequency divider in Figure 2.2 must provide a variable division ratio given in (2.3). For example, for a 10-MHz channel spacing, a crystal quartz of 10 MHz would make k be from 0 to M, instead of from 0 to $2M$ as in the case of a crystal quartz of 5 MHz. Moreover, for this last hypothesis, the odd harmonics of 5 MHz are also present in the signal path, which can result in errors of the control of the VCO.

$$N = N_L + k, \qquad k = 0, 1, \ldots, M \tag{2.3}$$

An example of this type of divider is the *pulse-swallow*, illustrated in Figure 2.3. This divider consists of a *prescaler*, a *program counter*, and a *swallow counter*. Before briefly describing the way it works three observations need to be made: (1) the prescaler divides the input by M or $M + 1$ depending on the logic state

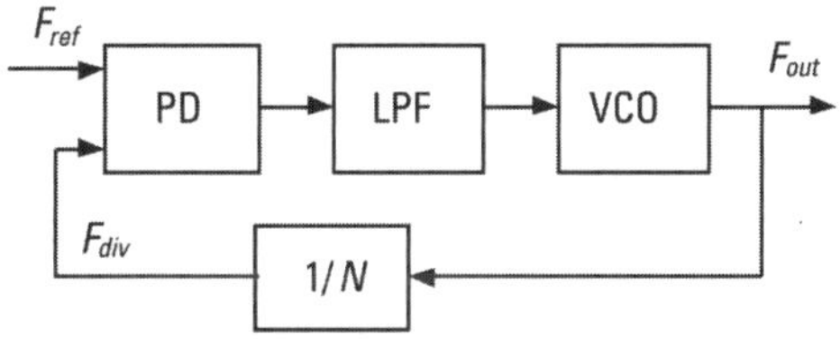

Figure 2.2 Integer architecture.

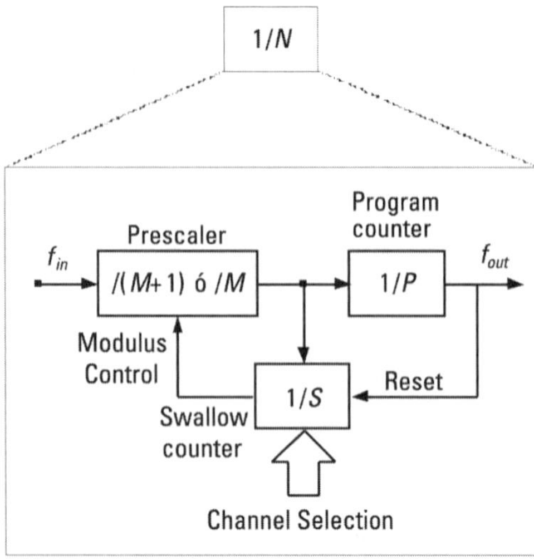

Figure 2.3 Pulse-swallow frequency divider.

of the control line of the module, (2) the program counter always divides the input of the prescaler by P, and (3) the swallow counter divides the gate of the prescaler by S, where S is determined by the digital channel selection input and can vary from one to the maximum number of channels. The swallow counter also has a reset input. The output frequency of this divider can be calculated using (2.4).

$$f_{out} = \frac{f_{in}}{MP + S} \tag{2.4}$$

When the circuit is in the reset state, the prescaler divides by $(M + 1)$. The output of this prescaler is divided as much by the program counter as by the swallow counter until this last one is complete; that is to say until S pulses have been counted. At this point, after $(M + 1) \cdot S$ cycles in the main input, the swallow counter changes the state of the module line control, making the prescaler divide f_{in} by M. Before this change is evident the program counter has already counted a total of S pulses. Following this change, the prescaler and the program counter keep on dividing until this last one is complete. Given that the program counter has already counted S pulses, $(P - S)$ cycles are required at its input and therefore $(P - S) \cdot M$ pulses at the main output in order to reach its end of count. Therefore, the main output generates a complete cycle every $((M + 1) \cdot S + (P - S) \cdot M)$ cycles at the input, that regrouping terms results in one every $(P \cdot M + S)$ cycles. The operation is repeated after the swallow counter is restarted.

The main characteristic of this architecture is its simplicity, which has made it a common choice in the implementation of integrated frequency synthesizers for many decades [3–12].

But this architecture also has some disadvantages. On the one hand, it is important to point out the appearance of spurious emissions in the output signal at a distance from the carrier equal to the reference frequency (F_{ref}), as illustrated in Figure 2.4.

These spurious emissions can be attenuated with an appropriate loop filter design. On the other hand, the reference frequency needs to coincide with the spacing between channels in order to be able to carry out the tuning of these. This requires a limitation in the loop bandwidth given that its determination is a trade-off between the time it takes to establish the channel (*lock time*) and, spurious emissions and phase noise at the output of the synthesizer. As the spurious emissions appear at a distance F_{ref} from the carrier, a bandwidth far superior to this frequency make these emissions attenuate little but also make the system quicker and vice versa. The same is true for the phase noise, a compromise between phase noise at the output and the speed of tuning is necessary for the determination of the bandwidth.

2.1.2 Fractional Architecture

As has been outlined the main disadvantage of integer-N architecture is that the reference frequency has to coincide with the separation between channels, which limits the bandwidth of the loop. On occasions the separation of channels is too small for integer-N architecture to be used making it necessary for the reference frequency to be higher than the said separation. On these occasions channel selection is carried out by varying the output frequency (F_{out}, see Figure 2.4) a fraction from the reference frequency (F_{ref}). This is the basis of fractional architecture synthesizers.

The fractional architecture block diagram is shown in Figure 2.5 [13]. This architecture substitutes the frequency divider from integer-N architecture

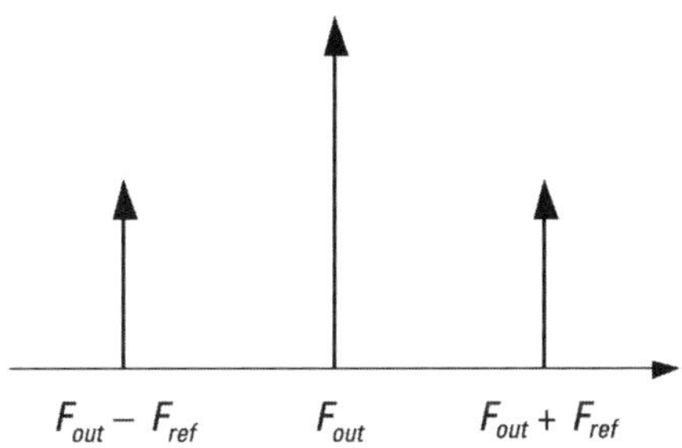

Figure 2.4 Spurious emissions due to reference frequency.

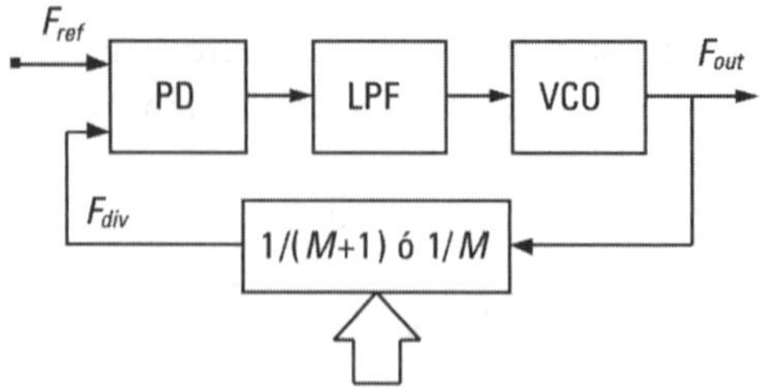

Figure 2.5 Fractional architecture using a dual-modulus divider.

for a dual-modulus prescaler. If A VCO output pulses are divided by this prescaler by the value M and B pulses by the value $(M + 1)$, then the ratio of division equivalent (N_{eq}) can be obtained from (2.5). This value can vary between M and $(M + 1)$ in fine steps using the correct choice of values A and B.

$$N_{eqt} = N \cdot f = \frac{A + B}{\frac{A}{M} + \frac{B}{M + 1}} \tag{2.5}$$

The resulting division ratio is sometimes denoted as $N \cdot f$, where the point refers to a decimal point and N and f represent the integer and fractional parts of the division ratio, respectively.

As previously mentioned, this type of architecture mitigates the need for the reference frequency to coincide with separation between channels. In this way, with F_{ref} in the range of tens of megahertz, the loop bandwidth of a fractional synthesizer can be in the order of just a few megahertz, something which produces a quick transient response together with a suppression of the VCO phase noise within this bandwidth. Furthermore, the decrease in the division ratios (when F_{ref} is increased) reduce the effect of the PD phase noise.

But this architecture presents the critical disadvantage of fractional spurious emissions in the same way that integer-N does. Fractional architecture presents spurs at fractional frequencies to those of the reference. For example, if the output frequency of the VCO is equal to $(N + \alpha) \cdot F_{ref}$, where N represents the integer part of the divisor and α the fractional part, the fractional spurs are introduced at frequencies $\alpha \cdot F_{ref}$, $2 \cdot \alpha \cdot F_{ref}$, and so forth. The problem with the fractional spurs is quite serious given that their levels are superior to those of the spurs that appear in integer-N architecture (typically between 20 and 30 dB below that of the carrier), the use of additional circuitry to compensate is usually necessary. These additional circuits can introduce considerable phase noise in the synthesizer output signal, and of course, add more circuit complexity.

2.2 Fundamental Figures of Merit of a Frequency Synthesizer

The design of a frequency synthesizer is a complex process in which a multitude of variables that can affect the correct working of the system exist. However, in the preliminary phases of design it is preferable to limit the number of variables in a way that initial decisions such as the architecture of each block or the technology can be adopted in an efficient manner. Therefore, in this section, three fundamental figures of merit (FOM) are considered: phase noise, spurious emissions and lock time.

Phase noise is a phenomenon that drastically affects the behavior of *transceivers*. In modern systems based on OFDM modulation it causes interference between subcarriers (ICI) as well as interference in the adjacent channel [14]. OFDM modulation divides each carrier in various orthogonally modulated subcarriers. As a consequence, and in the presence of phase noise, an extra ICI is added to the interference in the adjacent channel and which constitutes one of the key parameters when specifying the synthesizer phase noise requirement, as it appears as much outside as inside the loop bandwidth.

The same thing occurs with the level of spurious emissions but instead of affecting a frequency range like phase noise does, its influence is more specific.

The lock time of a synthesizer is also an important aspect that needs to be taken into account. When the modulus control selection input orders a change of channel, the synthesizer requires a finite time to establish the new frequency, or at least when it is settled to a few ppm with respect to the target value. Therefore, the lock time is an indicator of the speed of the system. This parameter is especially important in systems that employ the FHSS access technique.

Given the importance of these three aspects in the behavior of a frequency synthesizer, it is necessary to be able to estimate them very precisely before fabricating the device. Accurate specifications are the key factors in designing integrated circuits due to the high cost of prototyping. The following sections focus on two areas; the basic concepts and the estimation models cited in the references for designing CMOS outlining its main advantages and disadvantages.

2.2.1 Phase Noise

In order to estimate the contribution to the total phase noise of each one of the PLL blocks, a study based on the transfer function of each noise contribution needs to be carried out [15, 16]. To calculate these transfer functions, refer to the block diagram in Figure 2.6, where K_{ϕ} is the value of the charge pump current in A, K_{VCO} is the gain of the VCO in Hz/V, N and R are the modules of the main and the reference dividers, respectively, and $Z(s)$ the transfer function of the loop filter.

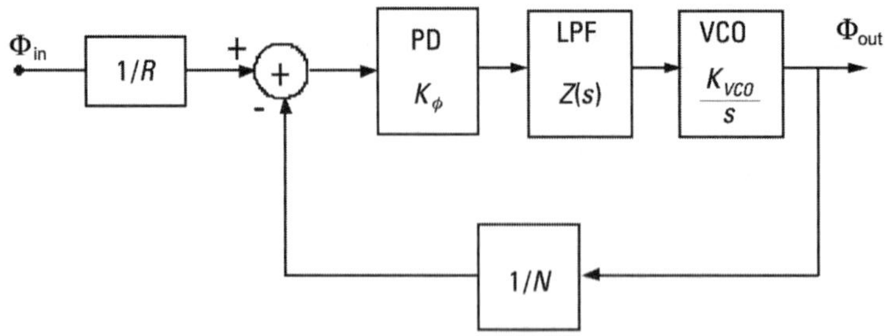

Figure 2.6 Diagram used to calculate the noise transfer functions.

The sources of noise are modeled as additive sources just after each one of the blocks and using basic control theory. As an example, Figure 2.7 shows the block diagram for the VCO noise (l_{VCO}).

Considering the PLL as a feedback system in the s-domain, it is useful to define the open-loop transfer function $G(s)$ (2.6) and the loop reverse gain $H(s)$ (2.7).

$$G(s) = \frac{K_\phi \cdot K_{VCO} \cdot Z(s)}{s} \tag{2.6}$$

$$H(s) = \frac{1}{N} \tag{2.7}$$

The closed loop gain can be calculated using the previous expressions:

$$F(s) = \frac{G(s)}{1 + H(s)\,G(s)} \tag{2.8}$$

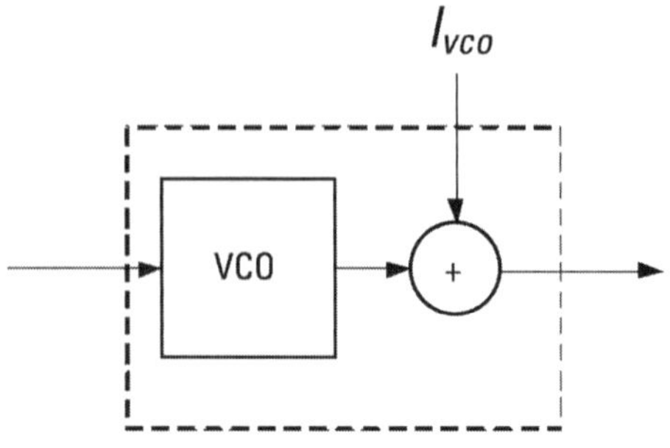

Figure 2.7 Diagram used to compute the noise contributions of each block.

The noise transfer functions of each one of the lock loop blocks can be derived finding the ratio between the response at the output and the corresponding input contribution. The functions obtained following this methodology are presented in Table 2.1. For the sake of brevity the calculations are not included here, but can be found in [17], where $T(s)$ is the transfer function of the noise voltage from the resistance and from the active circuit of the filter loop in the VCO. Obviously, the factor $\sqrt{2}$ has been introduced to convert the voltage from the efficient noise into peak voltage and factor $K_{VCO}/2f$, to transform this noise voltage into single side band (SSB) phase noise.

The function $T(s)$ is specific to the loop filter employed. In most integrated PLLs for high frequency applications this filter is passive; this is why the functions $T(s)$ corresponding to second- and third-order passive filters are presented in Table 2.2, where ZR_{21}, ZR_{22}, and ZR_{23} are given by (2.9), (2.10), and (2.11), respectively.

$$ZR_{21}(s) = \frac{\dfrac{C_2}{C_3}}{s(C_1 + C_2) + C_1 \cdot C_2 \cdot R_2 \cdot s^2} \tag{2.9}$$

$$ZR_{22}(s) = \frac{1 + R_2 \cdot C_2 \cdot s}{s(C_1 + C_2) + C_1 \cdot C_2 \cdot R_2 \cdot s^2} + R_3 \tag{2.10}$$

Table 2.1
Transfer Functions of the Main Sources of PLL Noise

Noise Source	Transfer Function
Reference crystal (l_{xtal})	$\frac{1}{R} \cdot \frac{G(s)}{1 + H \cdot G(s)}$
Divider R (l_R)	$\frac{G(s)}{1 + H \cdot G(s)}$
Divider N (l_N)	$\frac{G(s)}{1 + H \cdot G(s)}$
Phase detector (l_{PD})	$\frac{1}{K_\phi} \cdot \frac{G(s)}{1 + H \cdot G(s)}$
Loop filter (l_{LF})	$\frac{\sqrt{2} \cdot K_{VCO}}{2 \cdot f} \cdot \frac{T(s)}{1 + H \cdot G(s)}$
VCO (l_{VCO})	$\frac{1}{1 + H \cdot G(s)}$

Table 2.2
Transfer Functions $T(s)$ from Second- and Third-Order Passive Filters

	2° Order	3° Order
Circuit	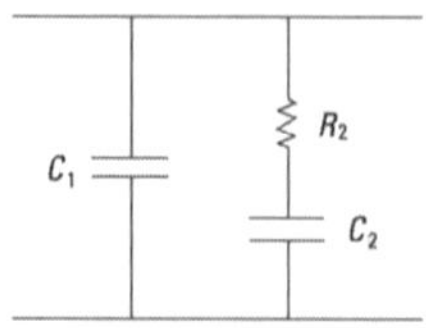	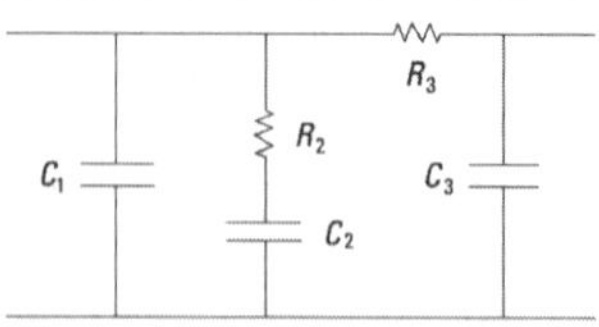
$T_{R2}(s)$	$\frac{C_2}{C_1 \cdot C_2 \cdot R_2 \cdot s + C_1 + C_2}$	$\frac{ZR_{21}}{ZR_{22} + ZR_{23}}$
$T_{R3}(s)$	—	$\frac{ZR_{23}}{ZR_{22} + ZR_{23}}$

$$ZR_{23}(s) = \frac{1}{s \cdot C_3} \tag{2.11}$$

The total phase noise at the PLL output is obtained by multiplying each source by its corresponding transfer function and adding all these resulting products. To transform this noise in dBc/Hz it is necessary to take the logarithm of the previous result, which is then multiplied by 10 or 20 depending on the units used. However, certain simplifications can be carried out in order to obtain a simpler expression for the design stage.

From Table 2.1 it can be seen that the frequency dividers, the reference crystal, and the phase detector present a common factor (which we will call $A(s)$) in their transfer functions given in (2.12), which tends towards N when $\omega \ll \omega_c$ and towards $G(s)$ when $\omega \gg \omega_c$. As $G(s)$ is a monotone function decreasing with frequency, these sources of noise are more influential within the bandwidth of the PLL.

$$A(s) = \frac{G(s)}{1 + H \cdot G(s)} \tag{2.12}$$

The loop filter and the VCO present the common factor given in (2.13). In contrast to the previous point this factor tends towards $N/G(s)$ when $\omega \ll \omega_c$ and towards 1 when $\omega \gg \omega_c$. Therefore, these sources of noise have more influence on offset frequencies superior to the band width of the PLL.

$$B(s) = \frac{1}{1 + H \cdot G(s)} \tag{2.13}$$

These considerations explain the graph presented in Figure 2.8 in which typical phase noise from a PLL is represented. It can be seen that at very small *offset* frequencies the source of noise that dominates is the reference crystal. Next, at slightly higher frequencies, but always smaller than the loop bandwidth, the phase detector noise dominates. Finally, at higher frequencies to this bandwidth, the biggest source of noise is the VCO with more or less influence on the loop filter, depending on what the actual level of noise is.

In most CMOS applications noise introduced by the two module frequency divisors R and N can be neglected (as it is discussed in Section 4.6). Therefore, using the transfer functions in Table 2.1 and neglecting the noise introduced by these dividers, the total phase noise at the output of the loop in dBc/Hz can be estimated using (2.14).

$$L\{\omega\} = 10 \cdot \log\Bigg[l_{xtal}(\omega) \cdot \left|\frac{A(s)}{R}\right|^2_{s=j\cdot\omega} + l_{VCO}(\omega) \cdot |B(s)|^2_{s=j\cdot\omega} + l_{PD} \cdot F_{ref}|A(s)|^2_{s=j\cdot\omega} + 2 \cdot k \cdot T \cdot R \cdot \frac{K_{VCO}^2}{f^2} \cdot |T(s) \cdot B(s)|^2_{s=j\cdot\omega}\Bigg] \quad (2.14)$$

where ω is the *offset* frequency with respect to the input frequency of the PLL, l_{xtal} and l_{VCO} the phase noise from the crystal reference and from the VCO,

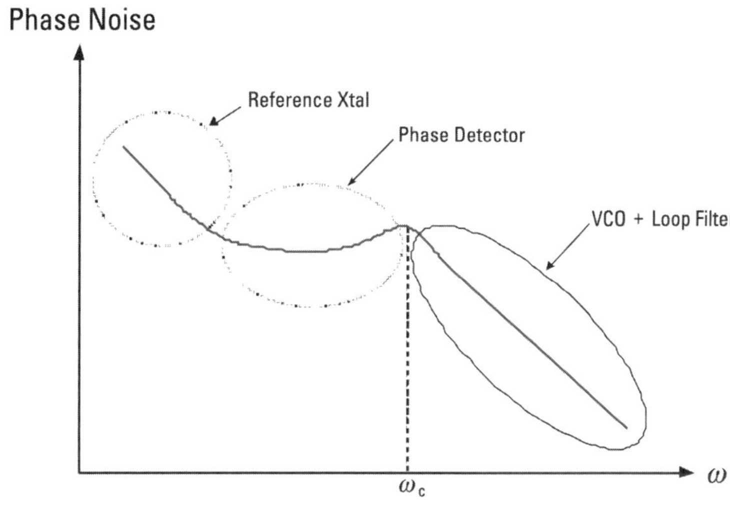

Figure 2.8 Typical phase noise from a PLL.

respectively, and l_{PD} the phase detector noise normalized at 1 Hz. To transform these last three noises into dBc/Hz, the logarithm needs to be taken and multiplied by 10. The study of each one of these terminals can be found in the following sections. Finally, the term corresponding to the loop filter makes a reference to a generic resistance of the filter, which is why a term for each resistance needs to be introduced, and in the case of using an active filter, another term that represents the noise of the active device.

2.2.2 Spurious Emissions

In Chapter 5, the nonideal effects that translate the phase detector into sources of spurious emissions are described. In addition, there are other distinct sources, which will be presented within this section.

There are no different design equations and simulation techniques for fractional synthesizers with respect to the integer-N ones. However, spurious emissions constitute a differentiating aspect in both types of synthesizers. In fractional synthesizers spurious emissions also appear at offset frequencies equal to the reference frequency, however as this frequency is usually much higher than that of integer synthesizers these emissions are strongly attenuated by the loop filter. But as well as these emissions, the spurious components that usually present considerable levels also appear at frequencies a fraction of the reference, which is why it makes it necessary to introduce a circuit to compensate them.

In order to understand the basic mechanisms that govern spurious emissions, the theoretical models of prediction that are presented throughout this section are focused mainly on integer architecture phase-locked loops.

In locked loop condition, a periodic jitter of the same frequency as the reference appears in the tuning path of the VCO. This effect gives rise to the so called reference spurs.

There are two fundamental causes that originate reference spurs: the *leakage* currents and the mismatches in the charge pump. In the next sections, each of these effects is described in more detail, presenting two models of prediction (Banerjee and Maxim models), and citing the conditions under which one or another type of spurious dominate.

2.2.2.1 The Banerjee Model

This model presents a detailed study of the two types of spurious emissions mentioned previously (leakage currents and CP mismatches), analyzing the main factors that give rise to each one and their impact on the overall reference spur problem. Furthermore, it suggests design equations that allow its levels to be estimated and analyzes the conditions that make one or another type of emission dominate. Finally, both types of spurious are combined in a single prediction equation [17]. The next points extract the key formulas needed to calculate these effects.

Spurious Based on Current Leakage (*Leakage-Spurs*)

At low reference frequencies (kHz), the effects caused by the leakage currents are the main cause of the reference spurious emissions. When the PLL is in a locked loop condition, the charge pump generates narrow pulses of currents spaced out by long periods of time, during which this charge pump is found to be in a high impedance state (*tristated*). However, this is just an ideal approximation, as the intrinsic characteristics of PN junctions provide leakage currents. These parasitic currents cause a modulation in the tuning path of the VCO producing the reference spurs.

To predict the level of reference spurious based on leakage in dBc, Banerjee proposes the use of the expression given in (2.15), where *BaseLeakageSpur* is a universal constant at an approximate value of 16 dBc applicable to whatever type of integer PLL. *Leakage* refers to the charge pump current loss and *SpurGain* is the gain of the spurious that can be calculated using (2.16).

$$LeakageSpur = BaseLeakageSpur + 20 \cdot \log\left|\frac{Leakage}{K_{\phi}}\right| + SpurGain \tag{2.15}$$

$$SpurGain = 20 \cdot \log\left|\frac{K_{VCO} \cdot K_{\phi} \cdot Z(s)}{s}\right|_{s=j \cdot 2 \cdot \pi \cdot F_{spur}} \tag{2.16}$$

where $Z(s)$ is the filter loop impedance and F_{spur} the *offset* frequency to which the spurious take place that in this instance is a multiple of F_{ref}.

Spurious Based on the Charge Pump Correction Pulses (*Pulse-Spurs*)

In classic PLLs literature it is completely normal to base reference spurs on leakage currents. For PLLs, where these currents are in the order of μA, this is a reasonable assumption. However, the most part of currently available synthesizers present current leakage in the order of 1 nA or less, and this is why other factors tend to dominate except when they are at very low reference frequencies.

The charge pump in a locked loop condition is inactive most of the time. It activates for short periods generating positive and negative current pulses that do not change the control voltage of the VCO but cause a jitter in its tuning path. It is these small pulses that generate the reference spurs (*Pulse-Spurs*). This type of reference spurs is also produced when trying to cancel the dead zone of the PFD (see Section 5.4).

The principal factors that affect the level of these reference spurs can be summed up in the following: the mismatching between the charge and the discharge currents of the charge pump, the difference between the activation times of the PMOS and NMOS transistors that act as switches in the charge pump, the dead zone elimination circuit, and the inaccuracies of the fractional calibration circuit. All these factors contribute to a variation in the width of the correction pulses of the charge pump in a locked loop state and influence the level of this type of reference spurious emission.

Note that the difference between the charge and the discharge currents from the charge pump is a function of its output voltage (i.e., the VCO control voltage). This difference, known as *mismatch*, is maximum in the extremes of the voltage range of VCO tuning. This is why the recommended tuning range of the VCO is the interval comprised between 0.5V and the charge pump voltage supply minus 0.5V. In this way the noticeable increase in the mismatch and the level of spurious is avoided.

Equation (2.17), proposed by Banerjee, estimates the level of these spurious in dBc as a function of the *SpurGain* (2.16), the frequency at which spurs appear F_{spur}, and the constant *BasePulseSpur*. In [17], a list of values for the constant *BasePulseSpur* corresponding to different frequency synthesizers, which range from −311 to −292 dBc, are presented. These values are linked to the examples presented and consequently have to be considered as an order of magnitude.

$$PulseSpur = BasePulseSpur + 40 \cdot \log\left(\frac{F_{spur}}{1\ \text{Hz}}\right) + SpurGain \qquad (2.17)$$

In this case, unlike what happens to the spurious based on leakage currents, the constant *BasePulseSpur* is not fixed, given that it depends on all the factors mentioned previously that influence the level of this type of spurs. In addition, it presents the disadvantage of being a purely empirical constant.

Combination of *Leakage-Spurs* and *Pulse-Spurs*

In most cases it is normally assumed that the total level of reference spurs is only due to *Pulse-Spurs*, but sometimes the level of the spurs based on leakage is also considerable. One way of determining what the type of dominant emission is consists of calculating the reference frequency for which the level of both types of spurs coincide. This can be achieved by using (2.18) and knowing beforehand the charge pump current leakage and the constant *BasePulseSpur*. If the real frequency reference is bigger than that calculated using (2.17), the *Pulse-Spurs* spurious emissions will be dominant and vice versa.

$$F_{spur} = 10^{\left[\frac{(BaseLeakageSpur - BasePulseSpur)}{40} + \frac{1}{2}\log\left|\frac{Leakage}{K_\phi}\right|\right]} \qquad (2.18)$$

Independently to the level of each one of the two types of spurious, the most exact way of predicting the total level of these emissions at the PLL gate is to add the contributions of each one of them. Equation (2.19) makes it possible for this estimation to be made.

$$Spur = 10 \cdot \log(10^{LeakageSpur/10} + 10^{PulseSpur/10}) \tag{2.19}$$

To conclude the presentation of this model it is interesting to point out that its principal advantage is the simplicity that it offers at the time of predicting the level of spurious. On the other hand, the most important disadvantage is that this prediction depends on eminently empirical constants. To overcome this disadvantage, the second model proposed by Maxim is now introduced.

2.2.2.2 The Maxim Model

According to this second model, the reference spurious emissions are fundamentally due to the fact that the charge pump is not ideal. In this way the principal factors that influence the level of these emissions are the leakage currents, the difference between the charge and discharge currents, and the temporary mismatching between the charge pump charge and discharge pulses in a locked loop condition [18].

According to Maxim, the level of reference spurs can be estimated using (2.20), where $\Delta\phi$ constitutes the phase error caused by the three factors mentioned earlier, BW is the bandwidth of the PLL in hertz, N is the transmission module unit divider and the feedback divider module, and f_p is the frequency of the extra pole added by the third-order passive filter. If the order is superior to three, the attenuation introduced by each one of the additional poles would need to be considered.

$$Spur = 20 \cdot \log\left(\frac{\Delta\phi \cdot N \cdot BW}{\sqrt{2} \cdot F_{ref}}\right) - 20 \cdot \log\left(\frac{F_{ref}}{f_p}\right) \tag{2.20}$$

The phase error $\Delta\phi$ consists of three terms that correspond with each one of the factors that give rise to the reference spurious emissions according to Maxim. This model proposes the estimation of this error using (2.21).

$$\Delta\phi = 2 \cdot \pi \cdot \left(\frac{Leakage}{K_\phi} + \frac{\Delta I}{K_\phi} \cdot \frac{T_{switch}}{T_{ref}} + \frac{\Delta T \cdot T_{switch}}{T_{ref}^2}\right) \tag{2.21}$$

where T_{ref} is the period of the reference signal at the PFD input, T_{switch} is the time during which the charge pump is found to be active in a loop locked condition, and ΔI and ΔT are the differences in current and time of the charge and discharge pulses of the pump in a loop locked condition.

2.2.2.3 Comparison Between the Banerjee and Maxim Models

The two models of predicting spurious discussed throughout this section are the model proposed by Banerjee and that of Maxim.

The former describes in great detail the causes that give rise to both types of spurs and provides numerous experimental results datum, such as a brief study of the conditions below which one or the other dominates. But it presents the main disadvantage of the use of eminently empirical constants, which are difficult to define for noncommercial synthesizers.

The second model distinguishes the same types of spurious as the first but it does not present such a detailed study. Its main advantage over Banerjee's model is that it does not use empirical constants to estimate the total level of the spurious at the loop gate. However, this model presents the disadvantage of basing its estimation on parameters of the synthesizer in a loop locked condition and this is why they cannot be precisely determined a priori either.

2.2.3 Lock Time

In addition to phase noise and reference spurs, the lock time of frequency synthesizer is also an important aspect to take into account. Generally, the time that a lock loop must use to change a gate frequency to another is limited and it is defined by the requirements of the standard of communication used. As a rule of thumb, this figure for the PLLs dedicated to RF communications can be estimated in the 0.1 to 1 ms range.

To arrive at an expression that allows the lock time of a PLL to be estimated it is necessary to use the closed loop transfer function $A(s)$ given in (2.12), which is defined as the ratio between the phase of the PLL gate signal and the phase of the reference signal. Once this transfer function has been calculated, it is assumed that the synthesizer gate signal changes from one frequency f_1 to another f_2, which is the same as the change in the reference frequency from f_1/N to f_2/N. Actually this jump in frequency in the reference signal can be modeled as a step function. In this way, calculating the Laplace transform of this step function and using the closed-loop transfer function, an expression of the PLL output frequency can be obtained in s-domain.

The time-domain function of the output frequency ($F_{out}(t)$) can be obtained applying the inverse Fourier transform to the previous expression (2.12). Equation (2.22) shows this result for PLL with a third-order passive loop filter, while expressions for a second-order system can be derived with $R_3 = C_3 = 0$.

$$F(t) = f_2 + (f_1 - f_2) \cdot e^{-\xi \cdot \omega_n \cdot t} \cdot \left[\cos\left(\omega_n \sqrt{1-\xi^2} \cdot t\right) + \frac{\xi - R_2 \cdot C_2 \cdot \omega_n}{\sqrt{1-\xi^2}} \cdot \sin\left(\omega_n \sqrt{1-\xi^2} \cdot t\right) \right] \tag{2.22}$$

where ω_n and ξ constitutes the natural frequency and the damping factor of the transient response of the locked loop, whose values can be calculated using (2.23) and (2.24), respectively.

$$\omega_n = \sqrt{\frac{K_\phi \cdot K_{VCO}}{N \cdot (C_1 + C_2 + C_3)}} \tag{2.23}$$

$$\zeta = \frac{R_2 \cdot C_2}{2} \cdot \omega_n \tag{2.24}$$

In Figure 2.9, the typical transient response of a locked loop synthesizer when its output frequency changes from one frequency f_1 to another f_2 is represented.

Finally, from (2.22) and considering an error in the final stabilization frequency of Δf, the lock time (T_L) of the synthesizer when it changes from one frequency f_1 to another f_2 can be calculated using (2.25).

$$T_L = \frac{-\ln\left(\frac{\Delta f}{f_2 - f_1} \cdot \frac{\sqrt{1-\xi^2}}{1 - 2 \cdot R_2 \cdot C_2 \cdot \xi \cdot \omega_n + R_2^2 \cdot C_2^2 \cdot \omega_n^2}\right)}{\xi \cdot \omega_n} \tag{2.25}$$

Note that (2.22) has been calculated simplifying the transfer function in locked loop $A(s)$ to another more simple second-order function.

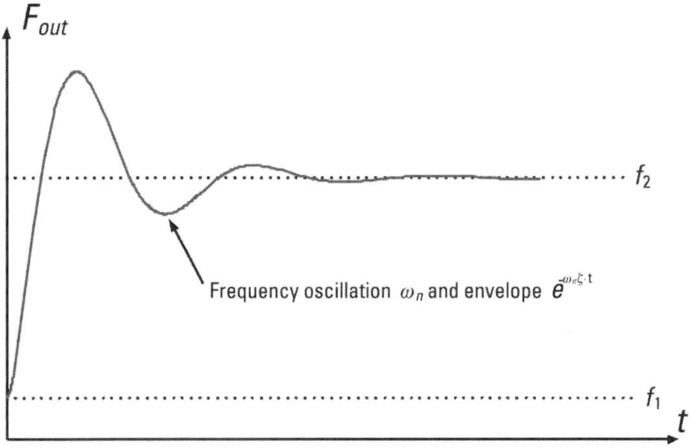

Figure 2.9 Typical transient response of a locked PLL for a step frequency change from f_1 to f_2.

References

[1] Craninckx, J., and M. Steyaert, *Wireless CMOS Frequency Synthesizer Design*, Norwell, MA: Kluwer Academic Publishers, 1998.

[2] Rohde, U. L., and D. P. Newkirk, *RF/Microwave Circuit Design for Wireless Applications*, New York: Wiley-Interscience, 2000.

[3] Lam, C., and B. Razavi, "A 2.6-GHz/5.2-GHz Frequency Synthesizer in 0.4-μm CMOS Technology," *IEEE Journal of Solid-State Circuits*, Vol. 35, No. 5, May 2000, pp. 788–794.

[4] Valla, M., et al., "A 72-mW CMOS 802.11a Direct Conversion Front-End with 3.5-dB NF and 200-KHz 1/f Noise Corner," *IEEE Journal of Solid-State Circuits*, Vol. 40, No. 4, April 2005, pp. 970–977.

[5] Herzel, F., G. Fischer, and H. Gustat, "An Integrated CMOS RF Synthesizer for 802.11a Wireless LAN," *IEEE Journal of Solid-State Circuits*, Vol. 38, No. 10, October 2003, pp. 1767–1770.

[6] Rategh, H., H. Samavati, and T. Lee, "A CMOS Frequency Synthesizer with an Injection-Locked Frequency Divider for a 5-GHz Wireless LAN Receiver," *IEEE Journal of Solid-State Circuits*, Vol. 35, No. 5, May 2000, pp. 780–787.

[7] Zhang, P., et al., "A 5-GHz Direct-Conversion CMOS Transceiver," *IEEE Journal of Solid-State Circuits*, Vol. 38, No. 12, December 2003, pp. 2232–2237.

[8] Behzad, A. R., et al., "A 5-GHz Direct-Conversion CMOS Transceiver Utilizing Automatic Frequency Control for the IEEE 802.11a Wireless LAN Standard," *IEEE Journal of Solid-State Circuits*, Vol. 38, No. 12, December 2003, pp. 2209–2220.

[9] Vassiliou, I., et al., "A Single-Chip Digitally Calibrated 5.15-5.825-GHz 0.18-μm CMOS Transceiver for 802.11a Wireless LAN," *IEEE Journal of Solid-State Circuits*, Vol. 38, No. 12, December 2003, pp. 2221–2229.

[10] Leung, G. C. T., and H. C. Luong, "A 1-V 5.2-GHz CMOS Synthesizer for WLAN Applications," *IEEE Journal of Solid-State Circuits*, Vol. 39, No. 11, November 2004, pp. 1873–1882.

[11] Zargari, M., et al., "A 5-GHz CMOS Transceiver for IEEE 802.11a Wireless LAN Systems," *IEEE Journal of Solid-State Circuits*, Vol. 37, No. 12, December 2002, pp. 1688–1694.

[12] Ahola, R., et al., "A Single-Chip CMOS Transceiver for 802.11a/b/g Wireless LANs," *IEEE Journal of Solid-State Circuits*, Vol. 39, No. 12, December 2004, pp. 2250–2258.

[13] Razavi, B., *RF Microelectronics*, Upper Saddle River, NJ: Prentice-Hall, 1998.

[14] Armada, A. G., "Understanding the Effects of Phase Noise in Orthogonal Frequency Division Multiplexing (OFDM)," *IEEE Trans. on Broadcasting*, Vol. 47, No. 2, June 2001, pp. 153–159.

[15] Lascari, L., "Accurate Phase Noise Prediction in PLL Frequency Synthesizers," *Applied Microwave & Wireless*, Vol. 12, No. 5, May 2000, pp. 329–330.

[16] Razavi, B., "A Study of Phase Noise in CMOS Oscillators," *IEEE Journal of Solid-State Circuits*, Vol. 31, No. 3, March 1996, pp. 331–343.

[17] Banerjee, D., *PLL Performance, Simulation, and Design*, 2nd ed., National Semiconductor, 2001.

[18] Maxim, A., "Low-Voltage CMOS Charge-Pump PLL Architecture for Low Jitter Operation," *Proc. Eur. Solid State Circuits Conf. (ESSCIRC)*, 2002, pp. 423–426.

3

LC-Tank Integrated Oscillators

As introduced in Chapter 1, the most suitable integrated frequency synthesizer for RF applications at working frequencies of various gigahertz is the phase-locked loop. This synthesizer includes an oscillator in its architecture, a component responsible for generating the synthesizer's output signal. The oscillator together with the divider is the block that presents the most difficulties at the time of design. This is mainly due to the fact that it is an analogical component with very high frequency of operation.

Concerning the architecture, the LC-tank is the most suitable for implementing the integrated oscillator using integrated inductors and varactors as passive elements.

These oscillators are suitable for full integrated implementation, and can also be operated at frequencies in the order of gigahertz with phase noise ranging around −115 dBc/Hz at 1 MHz for 5 GHz [1–12].

The objective of this chapter is to present the different architectures most commonly used to implement LC-tank oscillators whose resonant circuit is made up of integrated inductors and varactors, elements that are outlined in Section 3.3, followed by the models of phase noise found in the references. Finally, Section 3.5 outlines the most important considerations in the design of LC-tanks at the design layout stage.

Prior to the explanation on the diverse architectures found in recent integrated implementations (Section 3.2), their mode of operation is briefly described outlining the most important functioning characteristics.

3.1 Functional Description

As its own name indicates this type of oscillator is based on a LC-tank circuit that generates a periodic output signal when it resonates. The oscillator, understood as a single output circuit (Figure 3.1), is divided into the following three elements:

- Active circuit;
- Resonating circuit;
- Positive feedback path.

The ideal LC-tank resonant circuit determines the oscillation frequency at the output according to (3.1) and the active circuit is an amplifier that, by means of regenerative feedback, adds to the tank the energy needed to compensate its losses, and allow oscillation to be maintained.

If the inductor (L) and the capacitor (C) were ideal, that is to say without losses or any parasitic elements, a small signal would be sufficient for the oscillator to start to oscillate at the frequency (f_{osc}) given in (3.1).

$$f_{osc} = \frac{1}{2 \cdot \pi \cdot \sqrt{L \cdot C}} \tag{3.1}$$

However, as both the inductor and capacitor exhibit parasitic effects, the LC-tank presents signal loss or attenuation, as well as a shift in desired oscillation frequency (usually downward) caused by these parasitics. A conductance in parallel to the tank (G_p) can be used to represent the losses from the inductor and the capacitor, and the output resistance from the active circuit.

In Figure 3.2 both an ideal and real LC-tank are shown together with the necessary excitation so that they will start oscillating with their corresponding output signals. In this figure, the necessary excitation is represented by a current pulse. In real implementations, even small power resulting from the resistance noise is enough to provoke the oscillation of the output signal. Once oscillation has started, in the ideal scenario this state is indefinitely maintained. However, the oscillation in real LC-tanks is obviously dumped by internal losses in each cycle, unless the system is fed with extra power.

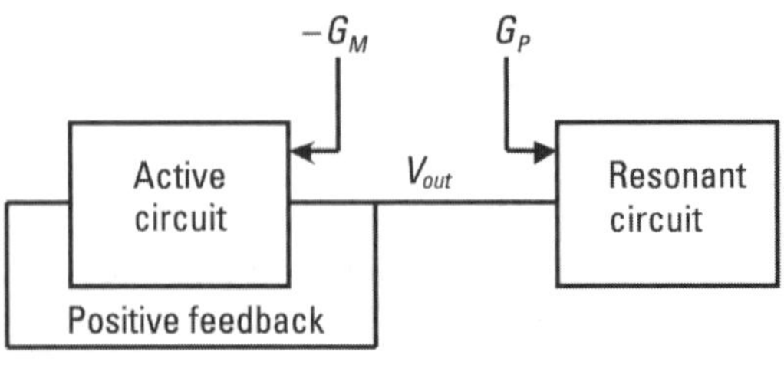

Figure 3.1 LC-tank simplified diagram.

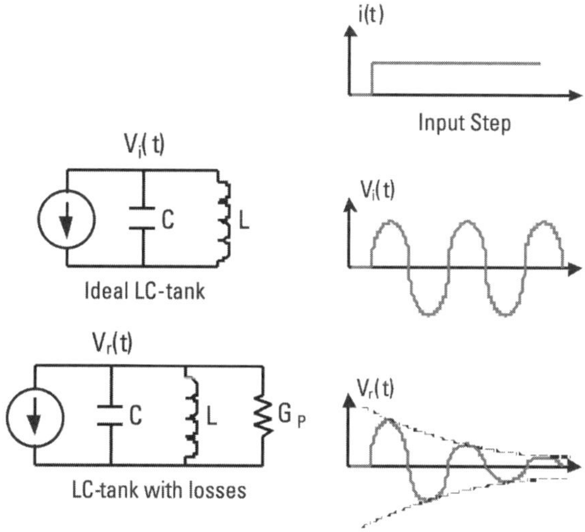

Figure 3.2 Ideal and real LC-tank transient output.

The different types of LC-tank oscillators differ basically in the way they generate negative resistance to compensate the losses in the tank. Equation (3.2) illustrates the condition that must be met to maintain oscillation in the tank indefinitely, where G_p is the conductance equivalent to the tank and G_M is the negative conductance generated by the active circuit.

$$|G_M| > |G_P| \tag{3.2}$$

If the conductance generated by the active circuit is chosen to be exactly the same as the tank conductance, the condition required for steady state oscillation would be fulfilled, but initial transient effects could prevent this state being reached. In order to guarantee the startup, it is recommended that the negative conductance is at least two or three times the positive conductance [13].

3.2 Types of LC-Tank Oscillators

Given that this book deals only with the design of PLLs for RF in CMOS technology, this section only shows the most commonly used configurations for implementing integrated LC-tank oscillators with CMOS technology.

Regarding the characteristic of the output signal, LC-tank oscillators can be classified into single-ended or differential. The principal advantages of the latter over the former are:

- High rejection of common mode interferers (e.g., noise coupled by substrate or supply sources, and effects of packaging);
- Better isolation;
- Strong attenuation of even order harmonics;
- Decrease of the dependence of the circuit on external variables such as temperature;
- Improvement of the quality of the chip's ground plane, reducing the influence of the bonding pads;
- It presents lower phase noise [14].

On the other hand, the fundamental disadvantages are:

- The need of approximately double the number of components, therefore the occupied area will also be nearly double;
- Power consumption is also nearly double.

Taking into consideration the type of transistor used for the implementation of the LC-tank, another alternative classification can be done: NMOS, PMOS, and CMOS [15–18]. The following sections describe each one of these in more detail.

3.2.1 NMOS

The active circuit of this configuration is formed by a pair of cross-coupled NMOS transistors. The main advantages of this configuration lie in its simplicity due to the reduced number of elements that form it, and consequently the noise that it contributes is minimal while providing high levels of linearity. It is suitable for low voltage VCOs.

But the clearest disadvantage is that its power consumption is relatively high. Due to the fact that the negative resistance is obtained with just two transistors, the bias current required to achieve this resistance is often too high for low-power-aimed devices.

The three main design alternatives to implement this circuit can be appreciated in Figure 3.3. There are two distinct ways of biasing the active circuit: with a current source (Figure 3.3(a, b)) and simply with a resistor (Figure 3.3(c)). In the configurations shown in (a) and (c) a capacitor in parallel with the current source is required to ensure that the source of the NMOS transistors is AC ground, given that this point is not directly connected to ground as occurs in configuration (b).

Configuration (c) is barely used to integrate oscillators because of the increased levels of noise over the other two, and the high sensitivity to external

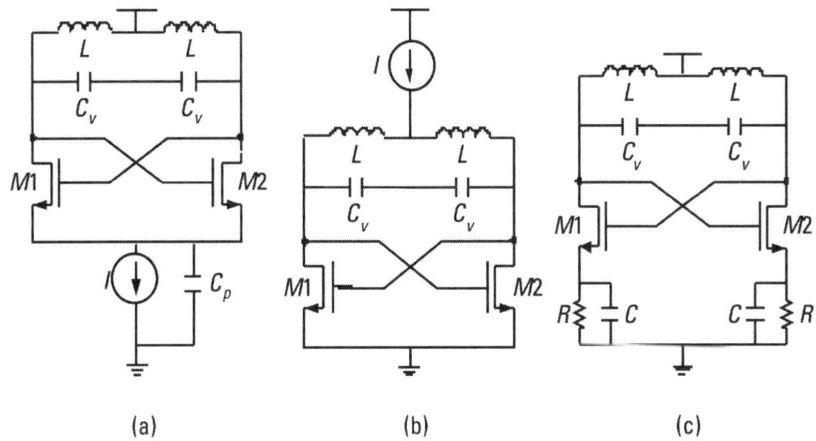

Figure 3.3 (a–c) NMOS differential LC-tank design alternatives.

factors such as changes in temperature, supply voltage, or dispersion in the fabrication process. However, it has the advantage of being the most simple of all three configurations, which is the reason why it is widely used in discrete realizations.

Concerning configurations (a) and (b), the selection of the most appropriate configuration for a particular case must take into account the limitations of the technology used. In the first case, the capacitor in parallel with the source of transmission usually requires a high value, which means also high chip area occupation. On the other hand, the second alternative does not require this capacitor, but PMOS transistors are necessary to implement the current source. This can also mean higher area occupation, given that the size of these transistors may be higher to obtain the same transconductance as with the NMOS transistors (see Section 1.1.1).

3.2.2 PMOS

The PMOS configuration is similar to the NMOS but using two PMOS type transistors instead of N type. The main problem lies in the usual worse performance of PMOS transistors compared to NMOS. Frequently PMOS require areas more that three times bigger that those used in NMOS in order to achieve the same negative resistance with the same power consumption, or a power consumption three times higher while maintaining the same area. As a result, the PMOS configuration is not regularly used, despite the fact that its transistors have a flicker noise superior to that of the NMOS.

3.2.3 CMOS

This configuration uses two N-type and two P-type MOS transistors to carry out the regeneration of the signal. With this the same amplification is achieved as in previous configurations (NMOS and PMOS) while requiring a lower current consumption. This solves the high current consumption drawback that the previous configurations present.

Among the disadvantages we can note, the need for a higher supply voltage required to bias the four transistors, the contribution of higher noise for the same bias current, the occupation of a higher area due to the higher circuit complexity, and a lower range of variation in the frequency of the signal. The origin of this final point has two sources. On the one hand, it is due to the increase in the fixed parasitic capacitance introduced by the two extra transistors and on the other hand the decrease in the range of control voltage in the tank that provokes these transistors.

The different forms of implementing the CMOS configuration are shown in Figure 3.4. The circuit diagram in Figure 3.4(c) presents the same disadvantages with respect to the other two in the NMOS case, which is why this diagram is not regularly used in integrated realizations.

As opposed to the previous case, Figure 3.4(b) requires a capacitor to create the AC ground condition for transistors M1 and M2. Taking into account that the current source occupies more space due to its implementation with

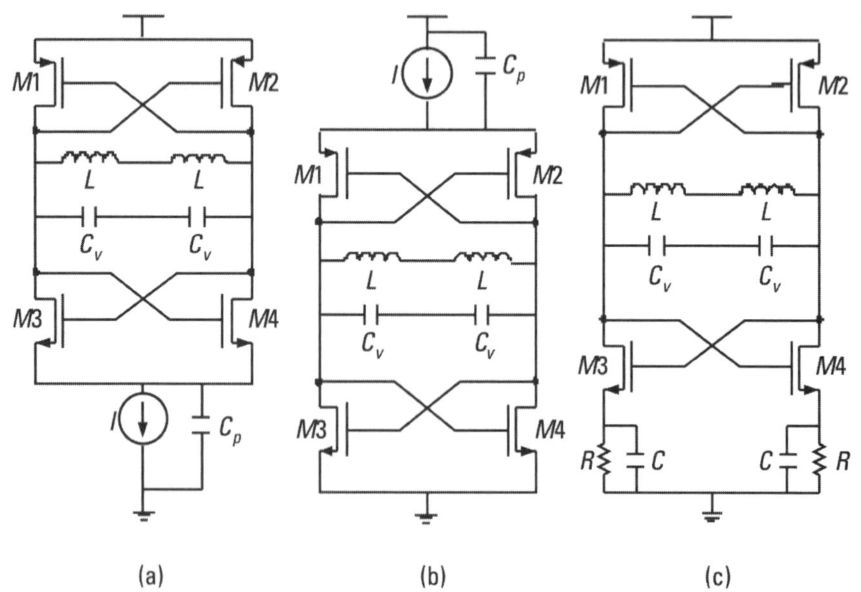

Figure 3.4 (a–c) CMOS differential LC-tank design alternatives.

PMOS transistors, the Figure 3.4(a) results are clearly more suitable for integrated implementation.

3.3 Integrated Passive Elements

The passive components are obviously key elements in LC-tank oscillators. The quality of both inductors and varactors determines the behavior of the whole system. In integrated circuits for high-frequency applications, the accurate design of the LC-tank is challenging because of several factors:

- For frequencies above 1 GHz, the parasitic effects of the rest of the components can be as representative as the actual values of the tank.
- Frequently the models of inductors and varactors provided by foundries have not been optimized for high frequency simulations.
- The quality of integrated inductors and varactors are usually well below the discrete realizations. Therefore, a clear optimization strategy must be used in order to obtain the best possible results.
- The range of real values that can be implemented with a particular technology is also limited compared to the discrete catalog.

These limitations make it crucial an in-depth study of the technology characteristics take place before the design of the PLL starts. In many instances, even the development of a complete library of passive components is worth the effort, because it will save on development time and will limit the number of unsuccessful fabrication runs.

3.3.1 Integrated Inductors

In Figure 3.5 a conventional integrated inductor in CMOS technology 0.18 μm is presented. The microphotograph shows the inductor with the test structure required to individually characterize the component.

In the following sections we are going to outline the most relevant ideas about models and different implementations. However, we want to stress that the objective of this book is the design of integrated PLLs, assuming that the designer has access to already developed models, and is familiar with main inductor configurations. If the reader feels the need to expand their knowledge on this particular point, we strongly recommend and encourage accessing further information using the specific references at the end of the chapter [19].

3.3.1.1 Π-Model

The electric model most commonly used to characterize conventional inductors is known as Π-model and its circuital diagram is presented in Figure 3.6.

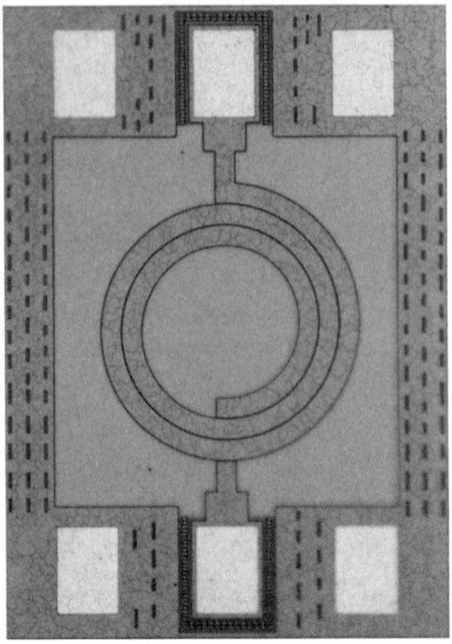

Figure 3.5 Integrated inductor.

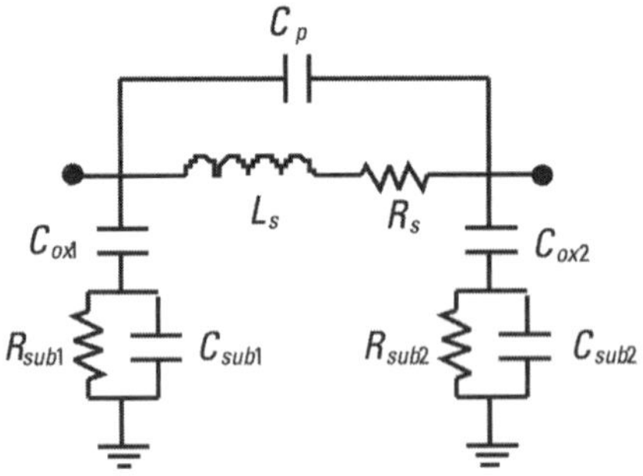

Figure 3.6 Π-model.

The physical meaning of each one of the ideal elements of this model is explained in the following points:

- C_p accounts for the capacitance between the metal tracks and these with the interior connection to the inductor.
- L_s represents the inductance of the inductor, including the self-inductance plus the mutual inductance between tracks of a same level and between tracks of different layers.
- R_s considers the resistance series of the inductor, including the ohmic losses of the metal tracks, the particular effects, and the losses due to the currents induced in the metal.

- C_{ox1} and C_{ox2} represent the capacities between the metal from the spiral and the substrate.
- R_{sub1} and R_{sub2} take into account the ohmic losses from the substrate produced by the currents induced magnetically and electrically in it.
- C_{sub1} and C_{sub2} modelize the capacitive effects of the substrate due to its semiconductor character.

The main advantages of this model lie in its simplicity, in the ease with which it can be adjusted to empirical results, and in the fact that it has a physical meaning, as is seen in each one of the elements that form it.

On the other hand, the main disadvantage of this model is its narrow bandwidth, that is to say it is only reliable in a frequency range of some hundreds of megahertz around its working frequency [19, 20]. This is due to the fact that nearly all its elements are frequency dependent. Despite this the Π-model is currently the most used for the characterization of integrated inductors.

3.3.1.2 Quality Factor

There are various definitions of the quality factor of an inductor (Q), but the most common is shown in (3.3), as the relation between the imaginary and real parts of the parameter y_{11}. This parameter is defined as the relation between the current and the voltage seen from output 1 of the inductor with output 2 connected to ground [21].

$$Q = \frac{|\mathrm{Im}(y_{11})|}{\mathrm{Re}(y_{11})} \tag{3.3}$$

As has been described earlier, the phase noise of a LC-tank oscillator depends a great deal on the quality of the passive elements that form the tank and especially on the Q of the inductor. Given that phase noise is one the critical parameters of PLLs, designing an inductor with the appropriate Q in order to fulfill the required specification of phase noise is a key factor for a successful design.

There are several ways to optimize the quality factor of an integrated inductor for a concrete inductance value. The first possibility is the optimization of the geometry of the inductor [18, 19, 22–24], and the second is the use of proper layout techniques for the area surrounding the inductor [25–28].

Another way to improve the performance is the use of certain technology features if the foundry gives us this opportunity. For example, some foundries offer top metal layers specially designed for the implementation of integrated inductors. They present higher thickness and higher distance from the substrate than the rest of the metallic layers intended just for interconnections. Another option is patterned ground shields using poly or metal layers as a means to help decouple the inductor from the substrate and hence reduce eddy currents that decrease L.

3.3.1.3 Inductor Configurations

In the design of differential devices, it is crucial to ensure the perfect symmetry of the main path of the signal. This fact adds another difficulty to the implementation of inductors, since the effect of input and output paths must be carefully taken into account in the models. Basically, there are two approaches to the design of symmetric LC-tanks. The first consists of using two conventional inductors in mirror configuration as shown in Figure 3.7.

The main disadvantage of this solution is that the occupied area may be high, which has a direct consequence on the die area occupation. Also, as a consequence of the extra resistance introduced by the metal path between both inductors, the quality factor can dramatically worsen.

To overcome those problems, balanced inductors can be chosen for the design [29]. This type of inductor presents identical behavior in each one of its outputs due to its geometric symmetry (Figure 3.8). In this way using an individual inductor would be enough in order to achieve symmetric behavior.

The balanced inductors are not suitable for designing a differential oscillator with NMOS architecture, due to the fact that in this type of architecture the power supply must be placed between the circuit tank's two inductors. Therefore, in this case the best solution is the use of conventional inductors in mirror configuration, while in order to implement an oscillator without this issue (for example, in CMOS architecture) balanced inductors are preferred.

3.3.2 Integrated Varactors

The capacitive component of an LC-tank oscillator can be integrated using different methods, but the most common is the implementation of integrated

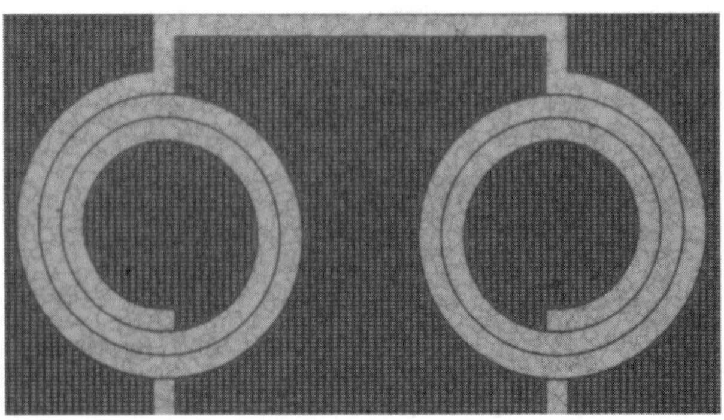

Figure 3.7 Micrograph of two mirror inductors for differential applications.

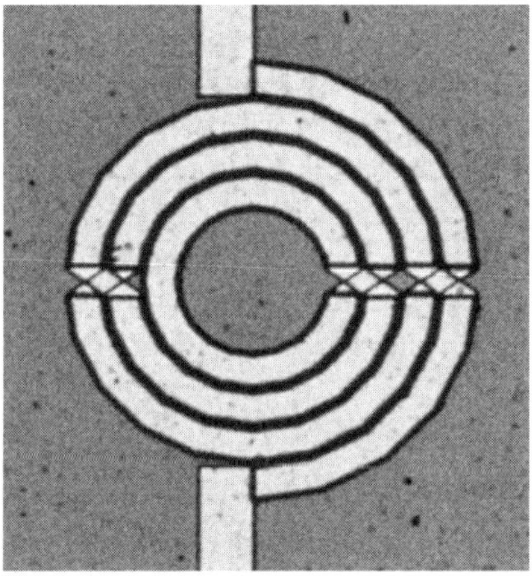

Figure 3.8 Microphotograph of a balanced inductor.

varactors because of its simplicity and its compatibility with CMOS integration processes. At an introductory level the varactors can be defined as variable capacitors tuned by means of an external voltage control and the parameters that characterize them are:

- The capacitance (C);
- The quality factor (Q), that can be roughly estimated using (3.4), where R is the resistance in series of the varactor and ω its operating frequency;
- The tuning range (γ), which expresses the variation in the capacitance of the varactor. It is calculated according to (3.5), where $C_{\max}$ is the maximum capacitance of the varactor and $C_{\min}$ is the minimum.

$$Q = \frac{1}{C \cdot R \cdot \omega} \tag{3.4}$$

$$\gamma = \pm \frac{C_{\max} - C_{\min}}{C_{\max} + C_{\min}} \tag{3.5}$$

Three basic types of integrated varactors can be named: PN junction varactors, varactors based on the MOS transistor, and triterminal varactors. The following briefly describes each one of them together with their advantages and disadvantages.

3.3.2.1 PN Junction Varactors

This type of varactor is mostly used in high-frequency designs due to its simplicity and stability. PN junction varactors are based on the variation of capacitance produced in the depletion zone of a PN junction when it is reverse-biased. In [30, 31] mathematic expressions can be found that allow for the approximate calculation of the capacitance of this type of varactor according to the reverse polarization voltage of the PN union. From these expressions it can be deduced that the variation of this capacitance with inverse polarization voltage is not linear, something which provokes the same nonlinearity in the variation of the oscillator output frequency with the voltage control. In [32] a simplified model is explained. It is based on the schematic circuit shown in Figure 3.9.

In this model:

- L_1 is the parasitic inductance of the P+ contacts.
- C_1 is the capacitance of the depletion zone between P+ diffusion and N-well.
- R_1 represents the P+ resistance.
- L_2 is the parasitic inductance of the N+ contacts.
- C_2 is the capacitance of the depletion zone between N+ buried layer and P– substrate.
- R_1 represents the N-well resistance.

The most relevant parameter is C_1, which can be calculated with the next expression:

$$C_1 = \frac{c_a A + c_w WL}{\left(1 + \frac{V}{V_0}\right)^{n1}} \tag{3.6}$$

where:

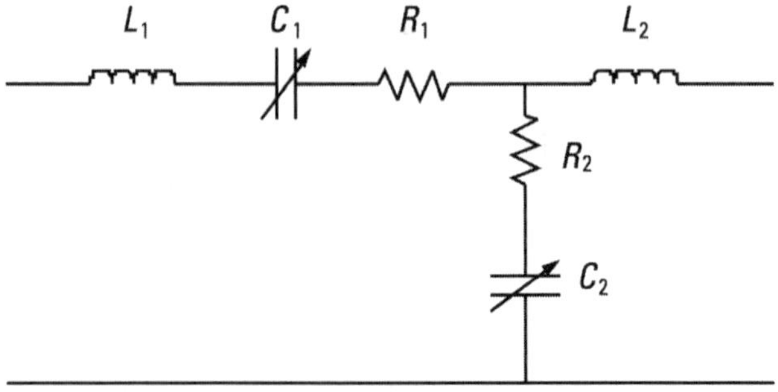

Figure 3.9 Simplified model of a PN junction varactor.

- A is the lateral area of the P+ diffusion and N-well.
- W, L are the transistor's width and length.
- V is the bias voltage.
- V_0 is the PN junction voltage.
- c_a is the capacitance per unit lateral area.
- c_w is the capacitance per unit base area.
- n_1 is a coefficient dependent on the doping level.

Figure 3.10 shows a typical graph of the variation of C_1 with respect to bias voltage.

With the objective of optimizing the characteristics of this type of varactor, it can be implemented using distinct geometries. The most typical are island varactors, matrix varactors, and finger varactors. This can be appreciated, respectively, in Figure 3.11(a–c).

The island varactor consists of a series of P+ and N+ islands placed in an N well. These islands are usually heavily doped to reduce the series resistance of the varactor, and as a result increase its quality factor.

In the case of the matrix varactor the N+ islands are substituted by an N+ mesh. With the P+ diffusions completely surrounded by N+ zones, the capacitance per area unit is maximized, at least compared to the island varactor. The main problem that it presents is the risk of avalanche breakdown, which can be avoided by careful inspection of minimum distances between the diffusions P+ and N+ in order to ensure that the separation is higher than the length of the zone of maximum depletion.

Regarding the finger varactor, it presents a higher tuning range than the two previous ones, (for the same occupied area of silicon as for the same capacitance of the varactor), despite suffering a slight reduction in the quality

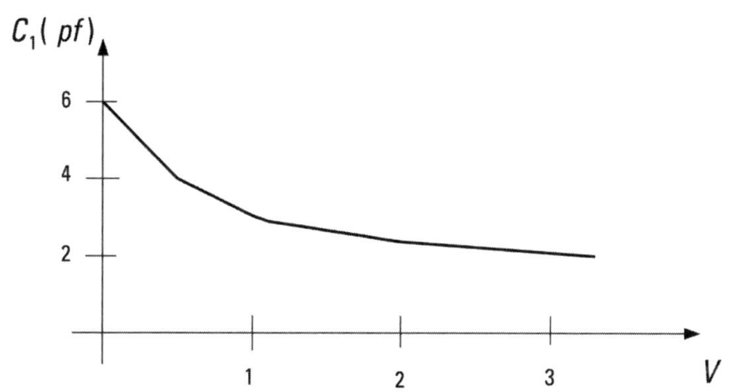

Figure 3.10 Typical C_1 versus bias voltage curve of a PN varactor.

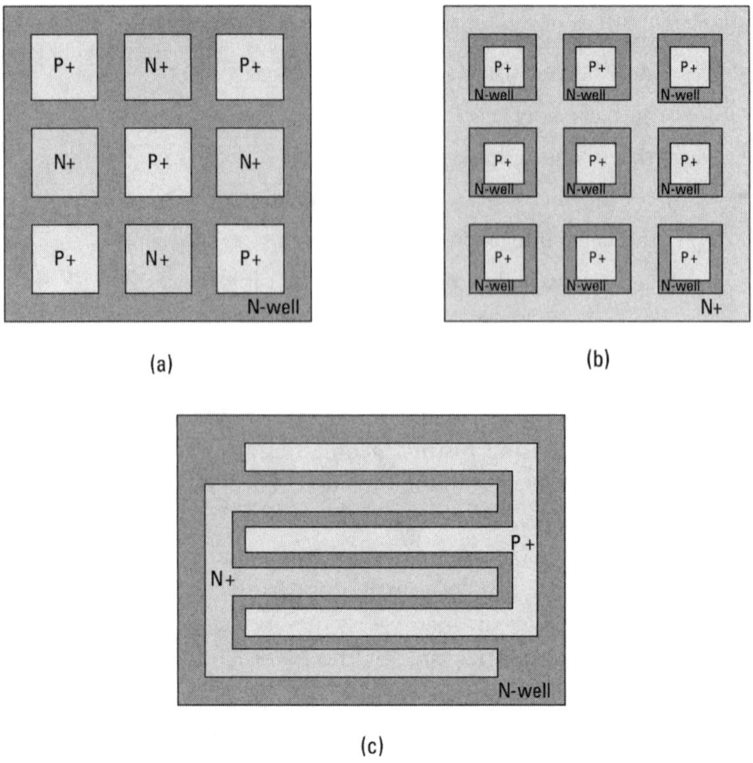

Figure 3.11 PN junction varactor geometries. (a) Island varactor, (b) matrix varactor, and (c) finger varactor.

factor principally with respect to island varactors. This reduction in quality factor has very little influence on the total quality factor of the tank, which is fundamentally marked by the quality factor of the integrated inductor. This is demonstrated with experimental measures in [32].

3.3.2.2 MOS Varactors

This type of varactor is probably the second most used after PN junction types. MOS varactors present a bigger capacitance and tuning range than those of a PN union, but in contrast, the variation in the capacitance is also more abrupt, something which can cause tuning problems in a VCO.

These varactors are based on the oxide capacitance at the output of a MOS transistor (fixed capacitance) together with the appearance of some parasitic capacitances in series or parallel that are variable with the voltage. The varactors operate in two modes: accumulation and inversion. Figure 3.12 shows a simplified schematic model.

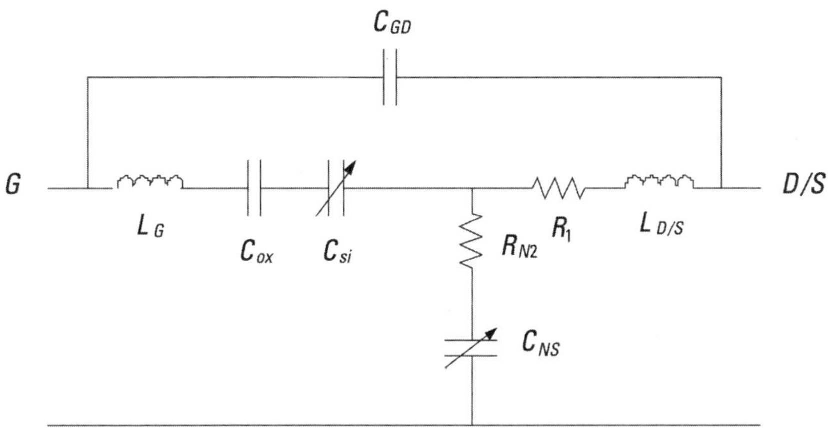

Figure 3.12 Simplified model of a MOS varactor.

In this model:

- L_G is the parasitic inductance of gate contacts.
- $L_{D/S}$ is the parasitic inductance of drain/source contacts.
- C_{si} is the capacitance associated to channel between drain and source.
- C_{ox} is the gate oxide capacitance.
- C_{GD} is the capacitance between gate and drain.
- C_{NS} is the capacitance between N+ buried layer and substrate.
- R_{N2} represents the N well resistance to the substrate currents.
- R_{N1} represents the N-well resistance to the gate to drain/source currents.

In this case, the most relevant parameter for understanding the working characteristics of these varactors is C_{si}, which can be calculated using the next expressions:

$$C_{si}(inversion) = \frac{c_i A_{ox}}{\left(1 + \frac{V}{V_{GSOFF}}\right)^n} \tag{3.7}$$

$$C_{si}(accumulation) = (c_a V + c_i) A_{ox} \tag{3.8}$$

In these expressions:

- A_{ox} is area of the junction formed by N-well and the oxide.
- V is the bias voltage.

- V_{GSOFF} is the voltage needed to remove the electrons of the N-well.
- c_i is a parameter for the inversion mode.
- c_a is a parameter for the accumulation mode.
- n is a coefficient dependent on technology parameters.

These varactors are based on the oxide capacitance at the output of a MOS transistor (fixed capacitance) together with the appearance of some parasitic capacitances in series or parallel that are variable with the voltage.

The different configurations of the MOS varactor result from the choice of the type of transistor used (NMOS or PMOS) and from the doping level of its drain/source diffusions (N+ or P+). The principal configurations are the PMOS, NMOS, N-accumulation and P-accumulation [32–35].

3.3.2.3 Triterminal Varactor

This type of varactor presents three terminals instead of two, obtaining an improvement in the tuning range with respect to the two previous.

Its working principle is a combination of the PN junction varactors and MOS varactors, so that the total capacitance seen from the drain has three components: the oxide capacitance of the output, the capacitance of the PN union, and some parasitic capacitances. Due to the presence of all these effects, a higher total capacitance is obtained per unit of area than with the other types of varactors. In addition, the variation of this capacitance is also wider thanks to the presence of the three terminals.

The operating regions are basically two, one in which the variable output voltage is applied to the gate maintaining the drain and the source at null potential, and the other in which the variation of the voltage is differentially applied to the drain and the source maintaining the gate at null potential [36].

As in the case of MOS varactors, the principal disadvantage of this varactor is the abrupt variation in its capacitance, something that can provoke problems of stability in a locked loop and excessive amplification of the noise at the output of the loop present in the tuning line of the VCO.

3.4 LC-Tank Oscillator Phase Noise

Phase noise is one of the most important parameters that characterize the functioning of an oscillator. This is why the specification of this figure of merit has a decisive influence on many design choices (e.g., the architecture of the oscillator, the tank circuit, and the active circuit). Therefore, the next sections are designed to present the different models most commonly used in order to evaluate the phase noise of a circuit during design stages.

In the frequency domain the models of Leeson and Craninckx stand out, both of which are based on the assumption that the oscillator is a linear system and temporally invariable (LTI). These models allow for a qualitative analysis of phase noise and enable us to obtain a mathematical expression for its estimation. The disadvantage of these models is that they do not take into account the conversion of flicker noise.

On the other hand, in the temporal domain the model of Hajimiri and Lee, which considers an oscillator as a linear system but variable in time (LTV), stands out, This model does take into account the noise introduced by the active circuit and the conversion of flicker noise, but it does not provide an expression that allows for the estimation of phase noise.

Before these models can be analyzed, the basic definition of phase noise is outlined.

3.4.1 Definition of Phase Noise

The output spectrum of a real oscillator consists of a fundamental tone at the frequency of oscillation plus several unwanted frequency components. When this undesired power is mixed with the output signal it produces lateral bands (phase noise) and harmonics as shown in Figure 3.13. In this figure the comparison between the spectrum of an ideal (single tone) and real oscillator can be seen.

Unwanted harmonics usually appear due to the lack of linearity in the elements that constitute the active circuit. As these are found relatively far from the main signal frequency, they often are nonrelevant in the context of a locked loop type synthesizer.

The output signal of a real oscillator according to time ($V_{out}(t)$) can be modeled using (3.9), where $A(t)$ and $\theta(t)$ are the amplitude and the phase from the output signal, respectively. Due to the fact that $A(t)$ like $\theta(t)$ are time-dependent functions, lateral bands (phase noise) appear in the spectrum

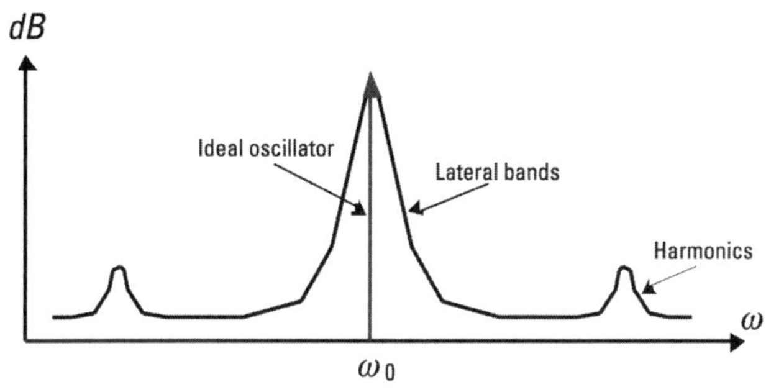

Figure 3.13 Ideal and real oscillator output spectrum.

of the oscillator output around its oscillation frequency ω_0. This is demonstrated in Figure 3.14.

$$V_{out}(t) = A(t) \cdot \sin[\omega_0 \cdot t + \theta(t)] \quad (3.9)$$

In order to measure phase noise, the power of noise in a 1-Hz bandwidth and at a distance of $\Delta\omega$ from the carrier is calculated and the result is divided by the power of the carrier. If the decimal logarithm is then calculated and the result multiplied by 10, the phase noise of an oscillator in dBc/Hz can be obtained as seen in (3.10) [37].

$$L\{\Delta\omega\} = 10 \cdot \log\left[\frac{P(\omega_0 + \Delta\omega)_{1Hz}}{P(\omega_0)}\right] \quad (3.10)$$

The impact of this negative effect on the receiver performance can be easily understood with the next figures. Let's assume that we would like to convert a frequency channel located at ω_1 to another frequency ω_2, and that neighbor channels stands at a distance of ω_{ad}, as can be seen in Figure 3.15.

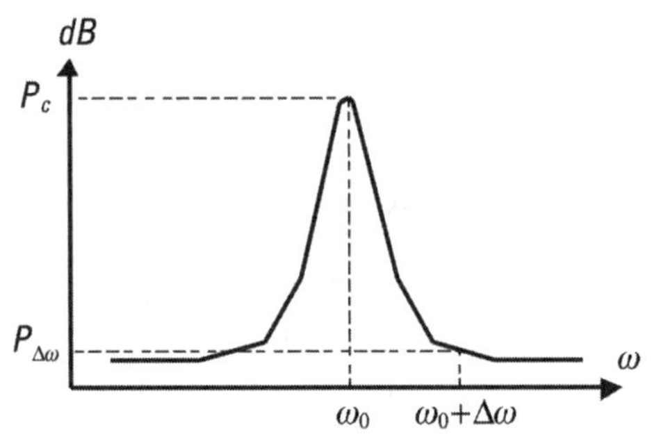

Figure 3.14 Typical output spectrum of an oscillator.

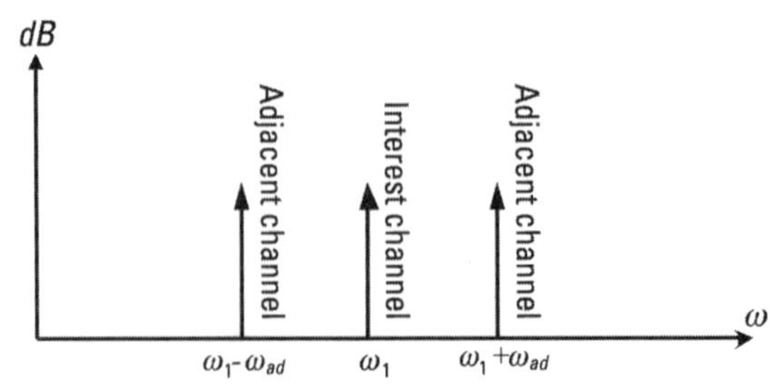

Figure 3.15 Channel distribution before frequency conversion.

In this discussion we are also assuming that the only nonideal source of the system is the local oscillator, as illustrated in Figure 3.16. In this graph it is straightforward to note the presence of lateral bands around the oscillation frequency, highlighting the power of noise at a distance ω_{ad} from the carrier.

After the frequency conversion, the channels exhibit the effect of phase noise, thus their spectrum spreads along the center frequency as shown in Figure 3.17. Therefore, the signal of interest receives unwanted signals that raise the noise floor of the channel.

This analysis of the influence of phase noise in the frequency translation has only considered interference from the adjacent channel. In OFDM systems the situation is far more complicated, as each carrier is divided into various subcarriers orthogonally modulated. The existence of these subcarriers generates extra interferences as a result of the nonideal effects of the local oscillator, called interference between subcarriers (ICI). This interference will be one of the key parameters at the time of specifying the requirement of phase noise from the synthesizer.

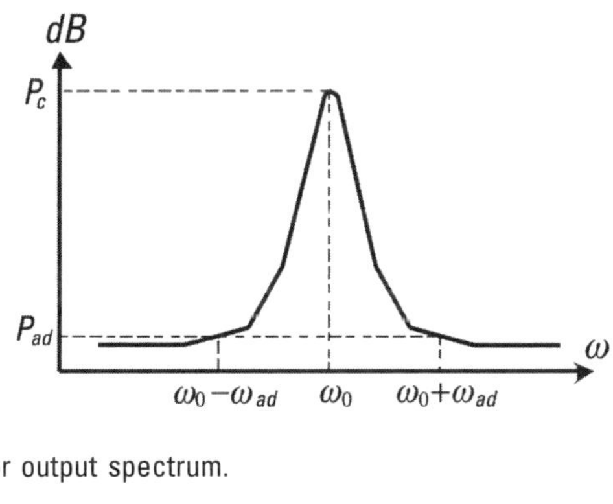

Figure 3.16 Oscillator output spectrum.

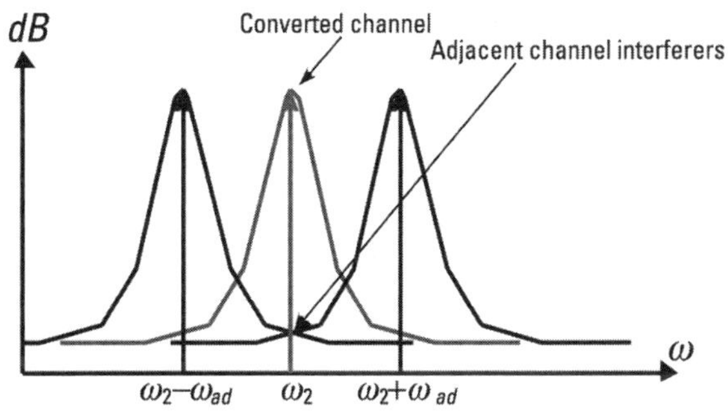

Figure 3.17 Channel distribution after frequency conversion.

Having defined both phase noise and its influence on frequency translation in a transceiver, the following sections present the main theoretical studies of this issue, both in the frequency and time domain.

3.4.2 The Leeson Model

This model of phase noise [38] carries out the analysis in the frequency domain and assumes that the oscillator is an LTI system. It provides an estimation of phase noise at the output of an oscillator as a combination of a theoretical base, which presents certain simplifications, and empirical modifications introduced to fit the final spectrum to the real case. The expression that allows for the estimation of phase noise in dBc/Hz from the oscillator according to this model is presented in (3.11).

$$L\{\omega\} = 10\log \cdot \left\{\frac{2 \cdot F \cdot k \cdot T}{P_{carrier}} \cdot \left[1 + \left(\frac{\omega_0}{2 \cdot Q \cdot \Delta\omega}\right)^2\right] \cdot \left(1 + \frac{\omega_{1/f^3}}{|\Delta\omega|}\right)\right\} \tag{3.11}$$

where k is the constant of Boltzman, T is the absolute temperature, Q is the quality factor of the LC-tank, ω_0 is the frequency of oscillation, $\Delta\omega$ is the offset frequency with respect to the carrier, $P_{carrier}$ the power of the carrier, F is an empirical factor that takes into account the increase in the noise density in the region $(1/\Delta\omega)^2$ (typical value is close to 2), and ω_{1/f^3} is the frequency that limits the regions $(1/\Delta\omega)^2$ and $(1/\Delta\omega)^3$ (Figure 3.18).

In (3.13) it can be seen that, according to this model, the phase noise from the oscillator decreases when the power of the output increases together with the quality factor of the tank, while the offset frequency remains constant with respect to the carrier ($\Delta\omega$).

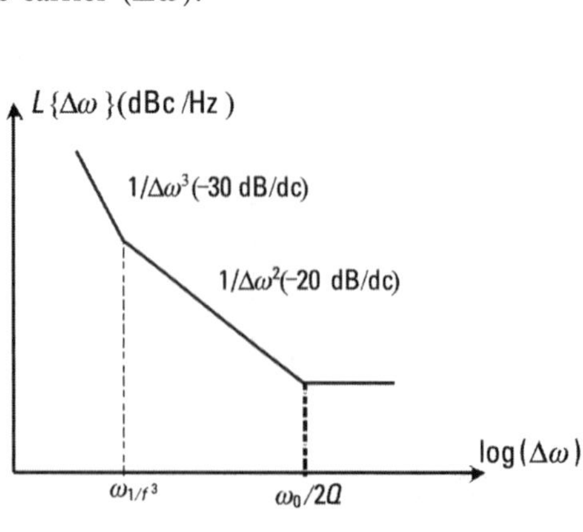

Figure 3.18 Leeson model graphical representation.

The main drawback of this model is that it presents an empirical constant F and a value ω_{1/f^3}, both difficult to calculate in design stages. In addition, according to [14], it doesn't provide a reliable prediction, only giving the designer a qualitative vision concerning the density of phase noise.

3.4.2.1 The Craninckx Model

This model is also based upon the assumption that the oscillator is an LTI system, but it also takes into account the noise introduced by the active circuit.

In Figure 3.19 the equivalent circuit to the oscillator is presented, with the sources of phase noise that have been considered in this model for the calculation of total phase noise.

The sources of noise that are considered in this model are the resistance series of the inductive component (L) from the tank $\left(\overline{v_{R_L}^2}\right)$, the resistance series of the capacitive component (C) of the tank $\left(\overline{v_{R_C}^2}\right)$, the resistance from the output of the active circuit and the parallel resistance of C and L $\left(\overline{i_{R_P}^2}\right)$, and the active circuit $\left(\overline{v_{G_M}^2}\right)$. In the Craninckx model [13], the contribution of each one of these sources of total noise from the output of the oscillator is calculated, in order to then finally combine them in a single expression given in (3.12).

$$L\{\omega\} = 10\log\cdot\left[2\cdot k\cdot T\cdot R_{eff}\cdot(1+A)\cdot\left(\frac{\omega_0}{V_a\cdot\Delta\omega}\right)^2\right] \tag{3.12}$$

where R_{eff} is the effective resistance from the tank, A is an empirical factor that models the excess of noise introduced by the amplifying of negative resistance,

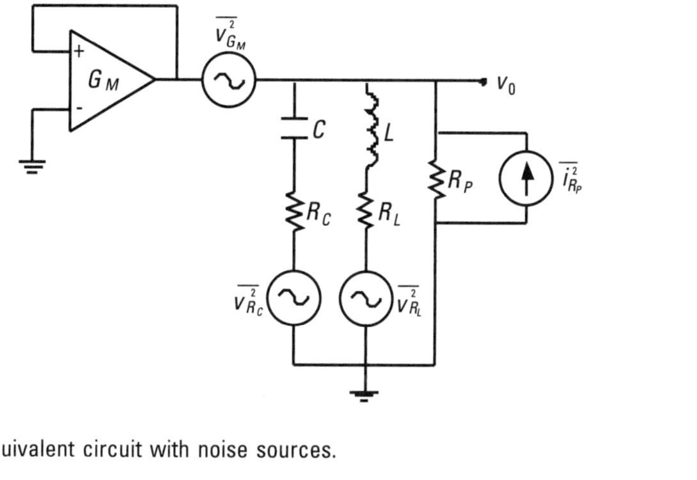

Figure 3.19 Equivalent circuit with noise sources.

and V_a is the amplitude of the periodic signal as seen from the output. The effective resistance from the tank is related to R_L, R_C, and R_P using (3.13).

$$R_{eff} = R_L + R_C + \frac{1}{R_P \cdot (\omega_0 \cdot C)^2} \tag{3.13}$$

Despite considering the noise that the active circuit introduces, the main disadvantage of this model lies again in the need to know an empirical factor a priori, which can be difficult to calculate in early stages of design.

3.4.2.2 The Hajimiri and Lee Model

This approach is more complex in comparison to the two previous models, however, its theoretical predictions adapt quite well to the characteristics of a real oscillator. This is why this model is used more in simulators than in estimated calculations carried out in the first stage of designing the oscillator [39].

This model assumes that the oscillator is a linear system variable in time (LTV) and studies the influence of total phase noise as much in active elements as in passive ones. Although the hypothesis of considering an oscillator as a linear system is not strictly true, in [37] it is shown that the effect of nonlinearity of an oscillator on its phase noise can be neglected. This paper also provides a really interesting overview across the LC-tank and the ring oscillators.

According to this model, the phase noise of an oscillator can be calculated through the variation on the phase of the output signal when an impulse function is applied at the input [40]. In order to clarify this calculation, Figure 3.20 shows an ideal oscillator, with an impulse current source at its input. This

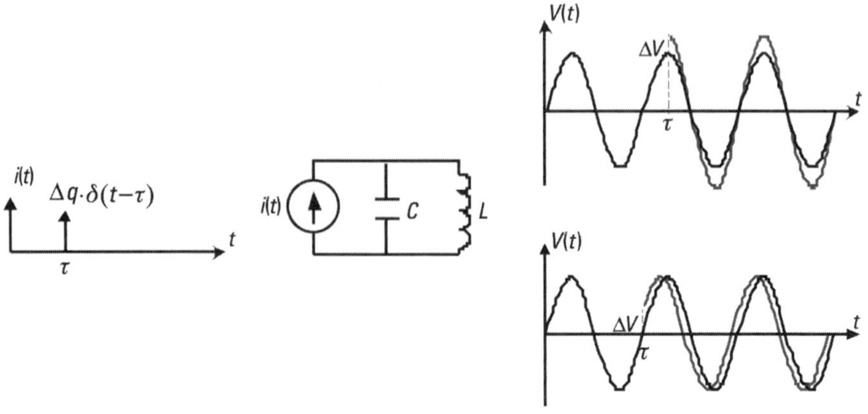

Figure 3.20 Ideal LC-tank output signal for impulse input.

source presents zero current for all the time values except in $t = \tau$, in which an area impulse Δq is produced. Given that the current in an inductor can not vary abruptly, the current impulse circulates through the capacitor, provoking an increase in the voltage of $\Delta V = \Delta q / C$.

As shown in Figure 3.20, the introduction of the current pulse can cause a change in the phase or in the amplitude of the output signal. If the impulse is applied when the output signal passes through zero, the change only affects the phase. On the other hand, if the impulse is introduced when the output signal passes through its maximum, the change only affects the amplitude of the signal. In both cases, the change affects the phase as much as the amplitude of the output signal.

As the system is linear its answer can be characterized according to the answer to the impulse function variable in time. The answer to an impulse of a system whose input is currents and the output is phase can be expressed using (3.14), where $q_{\max} = C \cdot V_{\max}$, $u(t)$ is the step function and $\Gamma(\omega \cdot t)$ is a periodic function of period 2π called impulse sensitivity function (ISF).

$$h_{\phi}(t, \tau) = \frac{\Gamma(\omega_0 \cdot \tau)}{q_{\max}} \cdot u(t - \tau) \tag{3.14}$$

The ISF function is an indicator of the sensitivity that a determined point presents when its phase varies in the presence of the impulse function. For the concrete case of an oscillator, the ISF is a cosine function.

From the impulse response of a LTV system it is possible to know the answer for any type of input from (3.15) [40], where $i(t)$ represents the noisy input current.

$$\phi(t) = \int_{-\infty}^{\infty} h_{\phi}(t, \tau) \cdot i(\tau)\, d\tau = \int_{-\infty}^{t} \frac{\Gamma(\omega_0 \cdot \tau)}{q_{\max}} \cdot i(\tau)\, d\tau \tag{3.15}$$

For a noisy current source of white noise with a density $\overline{i_n^2} / \Delta f$, using (3.15) the phase noise from an offset frequency from the carrier f_{off} can be derived in the following way:

$$L\{f_{off}\} = \frac{\Gamma_{rms}^2}{q_{\max}^2} \cdot \frac{\frac{\overline{i_n^2}}{\Delta f}}{16 \cdot \pi^2 \cdot f_{off}^2} \tag{3.16}$$

where Γ_{rms} is the average squared value of the ISF. When various sources of noise exist, $\overline{i_n^2} / \Delta f$ represents the total current in each node and is the sum in power of each one of the currents.

This model of phase noise, unlike the two previous models, includes the noise at low frequency known as flicker noise in its analysis, pink noise, or noise $1/f$. In reality, the sources of noise from the elements do not present white noise in all frequency components, but at low frequencies present an increase in noise that is the so-called flicker noise. This can be seen graphically in Figure 3.21, where $f_{1/f}$ represents the frequency of separation between these two types of noise.

The frequency conversion of flicker noise is determined by the DC value of the ISF. The ratio between $f_{1/f}$ and f_{1/f^3} (i.e., the frequency that separates the regions of different phase noise slope of −20 dB/dec and −30 dB/dec), can be obtained using (3.17), where c_0 is two times the dc value of the ISF.

$$f_{1/f^3} = f_{1/f} \cdot \frac{c_0^2}{2 \cdot \Gamma_{rms}^2} \tag{3.17}$$

As the amplitude of the positive and negative lobes of the ISF are determined by the flanks of the rise and fall of the output signal, the symmetry of these flanks reduce the value of c_0 and, as a consequence, attenuate the conversion of flicker noise.

It is important to point out that this model requires some excessively complex analytical calculations in order to be able to calculate the phase noise. As a consequence the deduction of an expression that provides the phase noise for a general case is not possible, but design recommendations are obtained from this model and it explains the procedure to follow in order to calculate this phase noise.

The main conclusions of this model obtained in [40] are the following:

- As long as the oscillator operates in the limited current region, an area of operation in which the output amplitude is inferior to the supply voltage, an increase in the current from the oscillator provokes an

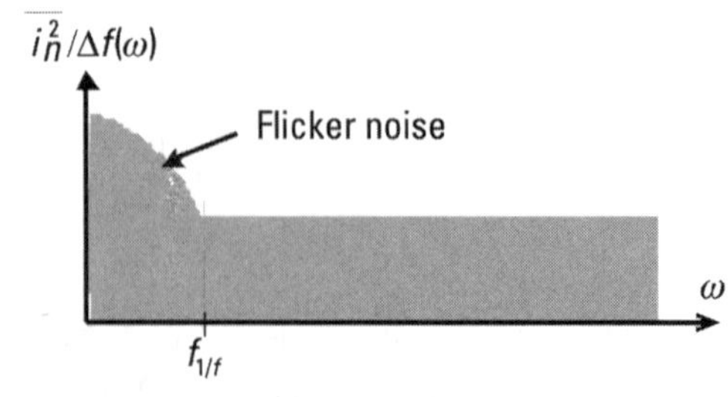

Figure 3.21 Flicker noise.

improvement in its phase noise as a consequence of the increase in the amplitude of the output signal.

- The phase noise depends on the supply voltage.
- Considering the two previous conclusions it can be deduced that in order to obtain the best phase noise possible for a given consumption of power, the current from the oscillator must be increased and the supply voltage decreased in the same proportion. Logically this is only possible as long as the supply voltage is sufficiently high to bias the transistors and to ensure the oscillator functions in the limited current region.
- Last, it is important to point out the final recommendation of maintaining the maximum symmetry possible in the design of the oscillator in order to avoid the conversion of flicker noise as far as possible.

3.5 Designing the Layout of the Oscillator

As it is always required in the design of high-frequency integrated circuits, the key point in order to obtain a reliable design is the accurate modeling of the undesired parasitic effects introduced by the components, metal layers, and contacts used to implement the schematic diagram. These parasitics are simple resistances, capacitances, and inductances added to the circuit. The designer must carefully check whether the simulation tool is considering all these parasitics, and in negative cases, their effects must be modeled with auxiliary components. Figure 3.22 shows the theoretical and generic parasitics configurations added to each combination of two nodes and a reference plane.

At these stages, some simple formulas can be used to estimate certain parameters, for example with (3.18) we can calculate the self-inductance of a

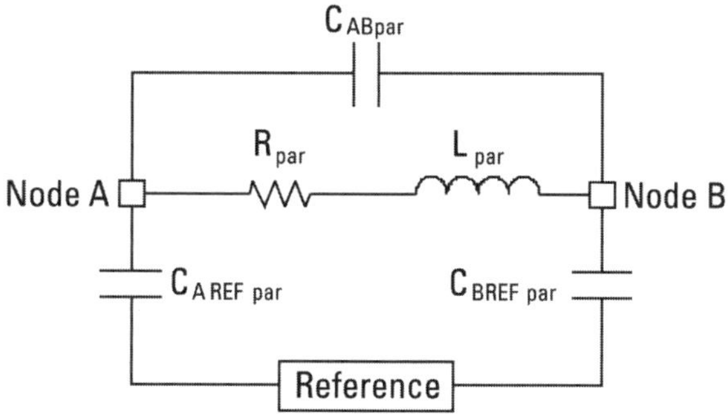

Figure 3.22 Parasitic equivalent circuit for a generic combination of two nodes and a reference plane.

rectangular conductor according to its length l, width w and thickness t in centimeters [41].

$$L = 2l \cdot \left[\ln\left(\frac{2l}{w + t}\right) + 0.5 + \frac{w + t}{3l} \right] \tag{3.18}$$

The appearance of extra parasitic resistances in the tank interconnections cause a worsening of the phase noise of the oscillator, because of the extra resistive thermal noise, and the need to increase the negative conductance provided by the active circuit in order to assure the correct functioning of the VCO.

In addition, the parasitic capacitances introduced in the *layout* cause a displacement in the frequency of oscillation and a reduction in the VCO tuning range. Finally, the most significant parasitic inductance is that caused by the track that connects the two inductors from the resonant tank in mirror configuration that equally provokes a displacement of the output frequency of the oscillator. Therefore, the key part of the layout that must always be carried out meticulously is the interconnections stage of the elements from the LC-tank. The metallic tracks used in these interconnections must be as short as possible with a suitable width depending on whether the introduction of less resistance or parasitic capacitance is desired. The wider these tracks, the less the parasitic resistance will be and vice versa.

Note with the passive inductive element, it is advisable that the interconnections of the elements from the tank are carried out by means of the top layer of metal. They usually have a lesser resistance and minimize the parasitic capacitance to the substrate because it is the furthest layer of metal from the substrate.

As an example, Figures 3.23 and 3.24 present the micrographs of two integrated oscillators [42, 43]. In Chapter 7 a more detailed design example is presented.

In these examples some important design features can be observed:

Isolation rings: Two large concentric rings surrounding the circuit are clearly illustrated. These are dedicated to the supply voltage and to the ground, rejecting any external influence that could damage or induce the slightest change in the VCO circuit, and obtaining a higher uniformity in the values V_{DD} and V_{SS}. In addition, surrounding the varactors, the inductors and transistors also have protection rings to ground in order to improve the isolation and reduce the noise to a minimum coupled through the substrate, with which the component of phase noise from external induction is minimized.

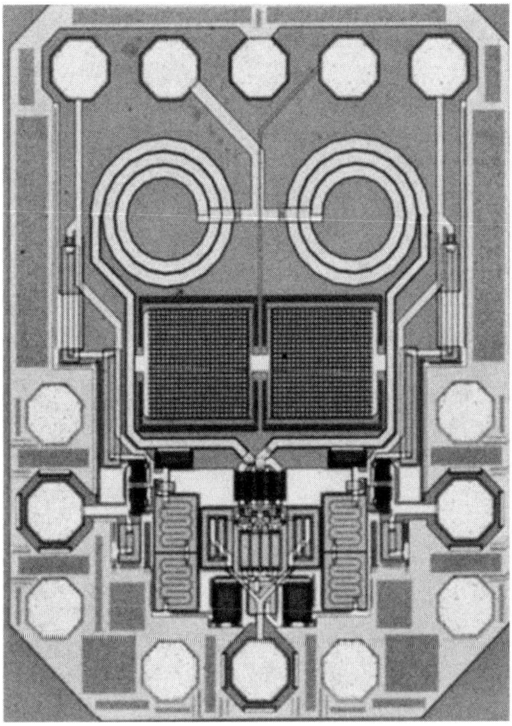

Figure 3.23 A 1.8-GHz SiGe 0.8 μm oscillator.

Ground contacts. They must be placed in the substrate throughout the circuit in order to reduce the effect of undesirable couplings. As previously outlined, all these ground contacts must be connected to the ground of the oscillator and connected to the ring that surrounds the circuit, but it should not be closed in order to avoid the appearance of parasitic inductive effects and substrate currents.

Decoupling capacitors: Capacitors in the order of tens of pF between the supply and ground voltages may be placed so that unwanted signals, due to the noise from the sources of voltage, find a low impedance path to ground. The rejection of these alternate components is important for the phase noise of the oscillator and of the complete PLL design.

Metallic tracks of interconnection between components: These must be designed using two criteria. First, these tracks must be capable of supporting the current that circulates through them, in this way avoiding electro migration effects that are one of the main causes of long-term failure in electronic circuits. For this reason it is imperative to follow the design rules recommended by each foundry basing maximum current values

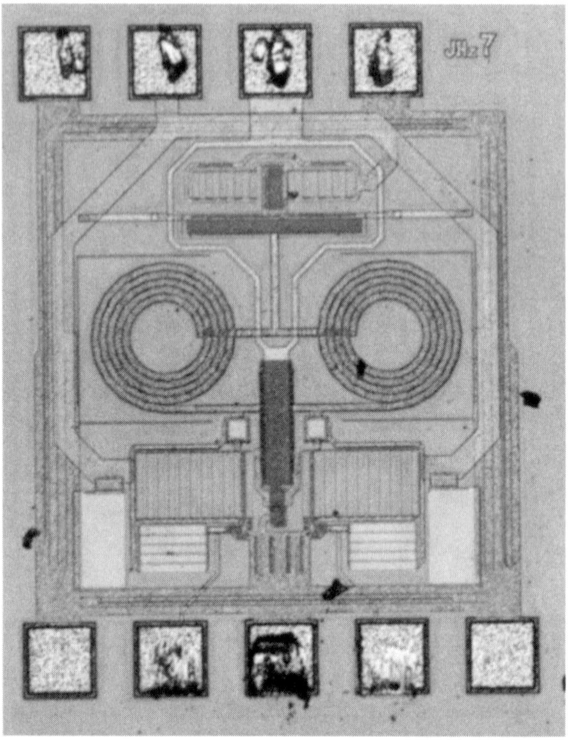

Figure 3.24 A 0.6-μm CMOS 1.56-GHz oscillator.

according to the width of the track. Second, to reduce the resistance, inductance, and parasitic capacitance introduced by the tracks as low as possible. Obviously, a trade-off between these targets must be carried out.

A key part of the design is the resonant tank and the transistors of the active circuit and of the output stage, and requires detailed planning. In Section 7.2.4, more details about this topic are provided.

As previously explained, the unavoidable amount of parasitic effects makes the redesign of the varactors essential once the layout of the oscillator has been completed and simulated in its postlayout version. Variations no higher than 5% should be registered but whose effect in the output parameters can turn out to be dramatic.

References

[1] Valla, M., et al., "A 72-mW CMOS 802.11a Direct Conversion Front-End with 3.5-dB NF and 200-KHz 1/f Noise Corner," *IEEE Journal of Solid-State Circuits*, Vol. 40, No. 4, April 2005, pp. 970–977.

[2] Zhang, P., et al., "A 5-GHz Direct-Conversion CMOS Transceiver," *IEEE Journal of Solid-State Circuits*, Vol. 38, No. 12, December 2003, pp. 2232–2237.

[3] Vassiliou, I., et al., "A Single-Chip Digitally Calibrated 5.15-5.825-GHz 0.18-μm CMOS Transceiver for 802.11a Wireless LAN," *IEEE Journal of Solid-State Circuits*, Vol. 38, No. 12, December 2003, pp. 2221–2229.

[4] Ahola, R., et al., "A Single-Chip CMOS Transceiver for 802.11a/b/g Wireless LANs," *IEEE Journal of Solid-State Circuits*, Vol. 39, No. 12, December 2004, pp. 2250–2258.

[5] Lam, C., and B. Razavi, "A 2.6-GHz/5.2-GHz Frequency Synthesizer in 0.4-μm CMOS Technology," *IEEE Journal of Solid-State Circuits*, Vol. 35, No. 5, May 2000, pp. 788–794.

[6] Bhattacharjee, J., et al., "A 5.8 GHz Fully Integrated Low Power Low Phase Noise CMOS LC VCO for WLAN Applications," *2002 IEEE MTT-S Int. Microwave Symp. Dig.*, Vol. 1, February 2002, pp. 585–588.

[7] Chu, Y., and H. Chuang, "A Fully Integrated 5.8 GHz U-NII Band 0.18-μm CMOS VCO," *IEEE Microwave and Wireless Components Letters*, Vol. 13, No. 7, July 2003, pp. 287–289.

[8] Hajimiri, A., and T. H. Lee, "Design Issues in CMOS Differential LC Oscillators," *IEEE Journal of Solid-State Circuits*, Vol. 34, No. 5, May 1999, pp. 717–724.

[9] Ham, D., and A. Hajimiri, "Concepts and Methods in Optimization of Integrated LC VCOs," *IEEE Journal of Solid-State Circuits*, Vol. 36, No. 6, June 2001, pp. 896–908.

[10] Zannnoth, M., et al., "A Fully Integrated VCO at 2 GHz," *IEEE Journal of Solid-State Circuits*, Vol. 33, No. 12, December 1998, pp. 1987–1991.

[11] Zargari, M., et al., "A Single-Chip Dual-Band Tri-Mode CMOS Transceiver for IEEE 802.11a/b/g Wireless LAN," *IEEE Journal of Solid-State Circuits*, Vol. 39, No. 12, December 2004, pp. 2239–2249.

[12] Zhang, P., et al., "A Single-Chip Dual-Band Direct Conversion IEEE 802.11a/b/g WLAN Transceiver in 0.18-μm CMOS," *IEEE Journal of Solid-State Circuits*, Vol. 40, No. 9, September 2005, pp. 1932–1939.

[13] Craninckx, J., and M. Steyaert, *Wireless CMOS Frequency Synthesizer Design*, Norwell, MA: Kluwer Academic Publishers, 1998.

[14] Hajimiri, A., and T. H. Lee, *The Design of Low Noise Oscillators*, Norwell, MA: Kluwer Academic Publishers, 2000.

[15] Hernández, J., et al., "Analysis of Architectures for 1.8 GHz CMOS LC-Tank Voltage-Controlled Oscillators," *Proceedings of DCIS'99*, Palma de Mallorca, November 1999, pp. 139–142.

[16] Egazi, E., H. Sjöland, and A. Abidi, "A Filtering Technique to Lower LC Oscillator Phase Noise," *IEEE Journal of Solid-State Circuits*, Vol. 36, No. 12, December 2001, pp. 1921–1930.

[17] Kucera, J. J., "Wideband BiCMOS VCO for GSM/UMTS Direct Conversion Receivers," *IEEE International Solid-State Circuits Conference*, Vol. 44, February 2001, pp. 374–375.

[18] Meléndez, J., "Diseño de un Triterminal de Conversión Directa a Baja Frecuencia Intermedia para GPS en tecnología CMOS," Doctoral Thesis, San Sebastián, 2001.

[19] Aguilera, J., and R. Berenguer, *Design and Test of Integrated Inductors for RF Applications*, New York: Springer, 2003.

[20] Niknejad, A. M., "Analysis, Design and Optimization of Spiral Inductors and Transformers for Si RF IC's," Proyecto Fin de Carrera, Berkeley, 1997.

[21] Ashby, K. B., et al., "High Q Inductors for Wireless Application in a Complementary Silicon Bipolar Process," *IEEE Journal of Solid-State Circuits*, Vol. 31, January 1996, pp. 4–9.

[22] Christensen, K., "Easy Simulation and Design of On-Chip Inductors in Standard CMOS Processes," Center for Integrated Electronics, Technical University of Denmark, 1999.

[23] Yue, C. P., and S. S. Wong "Physical Modeling of Spiral Inductors on Silicon," *IEEE Trans. on Electron Devices*, Vol. 47, March 2000, pp. 560–568.

[24] Razavi, B., *Basic Concepts in RF Design*, Lausanne, 2001.

[25] Ling Alan, P. L., "On-Chip Planar Spiral Inductor Induced Substrate Effects on RF IC's in CMOS Technology," Doctoral thesis, Hong Kong, January 1998.

[26] Hernández, E., "Integración de un conversor de frecuencia para TV en Tecnología SiGe 0.8 μm," Doctoral thesis, San Sebastián, 2002.

[27] Park, J. Y., and M. G. Allen, "Packaging-Compatible High Q Microinductors and Microfilters for Wireless Applications," *IEEE Trans. on Components, Packaging and Manufacturing Technology, IEEE CPMT-Part B*, March 1998.

[28] Kim, H. S., et al., "Spiral Inductors on Si p/p+ Substrates with Resonant Frequency of 20 GHz," *IEEE Electron Device Letters*, Vol. 22, No. 6, June 2001, pp 275–277.

[29] Long, J. R., and M. A. Copeland, "The Modeling, Characterization and Design of Monolithic Inductors for Silicon RF IC's," *IEEE Journal of Solid-State Circuits*, Vol. 32, March 1997, pp. 357–369.

[30] Neudeck, G. W., and R. F. Pierret, *The PN Junction Diode*, Reading, MA: Addison-Wesley, 1989.

[31] Pedersen, E., "RF CMOS Varactors for Wireless Applications," Doctoral thesis, Aalborg (Dinamarca), 2001.

[32] Gutiérrez García, I., "Varactores Integrados de Alto Factor de Calidad en Tecnología Estándar 0.8 μm SiGe Para Aplicaciones en RF," Doctoral thesis, San Sebastián, 2004.

[33] Andreani, P., and S. Mattison, "On the Use of MOS Varactors in RF VCO's," *IEEE Journal of Solid-State Circuits*, Vol. 35, No. 6, June 2000, pp. 905–910.

[34] Soorapanth, T., "Analysis and Optimization of Accumulation Mode Varactor for RF ICs," Center for Integrated Systems, Stanford University, California, October 1998.

[35] Shing, H. P., "High-Q and Wide Tunning Range Gated Varactor for Use in SOI and Bulk CMOS Radio Frequency Integrated Circuits," Doctoral thesis, Hong Kong University of Science and Technology, Hong Kong, December 1998.

[36] Gutierrez, I., J. Meléndez, and E. Hernández, *Design and Characterization of Integrated Varactors for RF Applications*, New York: Wiley, 2007.

[37] Lee, T. H., and A. Hajimiri, "Oscillator Phase Noise: A Tutorial," *IEEE Journal of Solid-State Circuits*, Vol. 34, No. 3, March 2000, pp. 326–336.

[38] Leeson, D. B., "A Simple Model of Feedback Oscillator Noise Spectrum," *Proc. IEEE L.*, Vol. 54, February 1966, pp. 329–330.

[39] Huang, Q., "Circuit Design for Wireless Communications," *Curso de RF*, Lausanne, Suiza, June 2001.

[40] Hajimiri, A., and T. H. Lee, "A General Theory of Phase Noise in Electrical Oscillators," *IEEE Journal of Solid-State Circuits*, Vol. 33, No. 2, February 1998, pp. 179–194.

[41] Grover, F. W., *Inductance Calculations: Working Formulas and Tables*, New York: D. Van Nostrand Co., 1962.

[42] Hernandez, E., et al., "Fully Integrated CMOS VCO for DCS 1800 Direct Conversion Receivers," *Microwave Journal*, September 2003, pp. 182–192.

[43] Hernandez, J., R. Berenguer, and G. Bistue, "Basic Guidelines to Design RF CMOS Cells for Wireless Receivers," *Microwave Engineering Europe*, April 2001.

4

Frequency Divider

This chapter focuses on the frequency divider, another component of the PLL. This block, together with the oscillator, is the most difficult to implement due to its high working frequency.

This section presents the different architectures for frequency dividers, drawing comparisons between their operating frequency and consumption. In addition, a thorough theoretical study on phase noise from frequency dividers is also presented.

Finally, following the same scheme used in the previous chapter, the last section is dedicated to outline the main recommendations for the generation of the layout from the schematic circuit.

4.1 Basic Frequency Dividers

In its more basic implementation the frequency dividers are basically made up of logic gates and flip-flops. They can be grouped in synchronous and asynchronous types, depending on how the synchronization of these flip-flops is performed [1]. In the synchronous dividers, each flip-flop is triggered by the input signal of the divider (clock). On the other hand, in the asynchronous dividers the input signal of the divider feeds the first flip-flop, which triggers the second and so on. Therefore, the synchronous frequency dividers achieve a complete transition faster than the asynchronous ones.

Figure 4.1(a) demonstrates a three-stage asynchronous divider. Each stage consists of a divider by two, whose output is the input of the next stage. This makes the third stage asynchronous with respect to the input to the clock. If

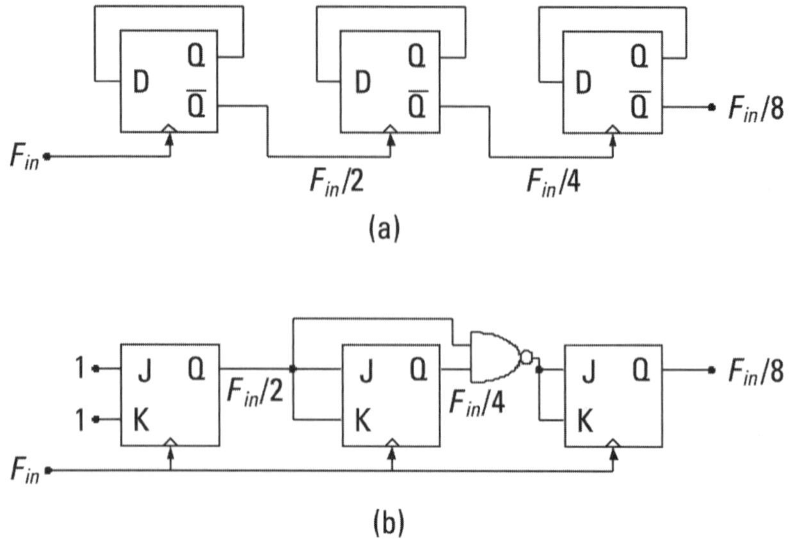

Figure 4.1 Frequency dividers (a) asynchronous and (b) synchronous.

the outputs of the three stages are combined together with the input in an AND gate, the output signal in this gate will be synchronized with the clock (with a small delay caused by the gate). The main disadvantage of the asynchronous dividers is the accumulation of the jitter (time-equivalent parameter of phase noise) from one stage to the other. This can be reduced placing a synchronization flip-flop at the end of the chain. In this case the jitter from the output is only that generated by the synchronizer [2].

The synchronous dividers can be implemented using JK-flip-flops, as illustrated in the example given in Figure 4.1(b). Each stage changes almost simultaneously in the clock edges. This makes the last stage respond more quickly than in the case of the asynchronous divider shown in Figure 4.1(a). Special care needs to be taken when the intermediate outputs of each one of the stages are combined in any AND logic gate. This is due to the fact that undesired glitches may appear when any of the signals change more quickly or more slowly than the others.

Although these basics could be used to achieve many different frequency reductions, the basic frequency divider may suffer from limitations for high-frequency operations due to the presence of PMOS transistors and to their digital focused purpose. In those cases, the system must include more building blocks in order to accommodate the signal to the limitations of these dividers. In the next section we review the main architectures used for this operation, specifically designed for high-frequency operation.

4.2 High-Frequency Divider Architectures and Building Blocks

There are two main categories of high-frequency dividers: fixed or tunable. As their names indicate, they differ in the type of division performed in each case. An adequate selection would depend on the channels' selection requirements for the application objective, as seen in Section 2.1. Figure 4.2 presents the basic diagram of both alternatives.

The fixed-N divider is the simplest alternative as far as structure, dimension, and current consumption are concerned but it is not valid for numerous applications that require implementation of a channel selector. This alternative, in Figure 4.2(a), is composed of fixed dividers, which progressively reduce the VCO frequency down to a suitable value to allow the comparison with the reference crystal signal.

With regard to tunable dividers, the most commonly used configuration is the architecture known as pulse-swallow, which is composed of three blocks: prescaler, program counter, and swallow counter, as illustrated in Figure 4.3(a). By contrast, the structure shown in Figure 4.3(b) involves a more complex circuitry, including a sigma-delta modulator, but it also offers more flexibility to deal with different channel widths.

The difference between these two types of tunable dividers is the channel spacing that can be tuned by their control. In the first case in Figure 4.3(a),

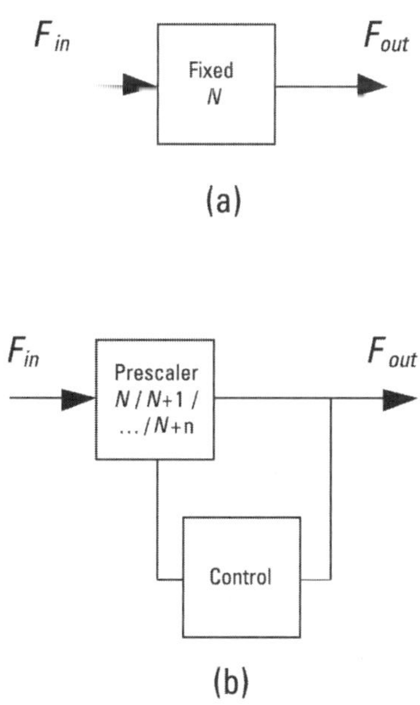

Figure 4.2 Frequency dividers: (a) fixed-N and (b) tunable.

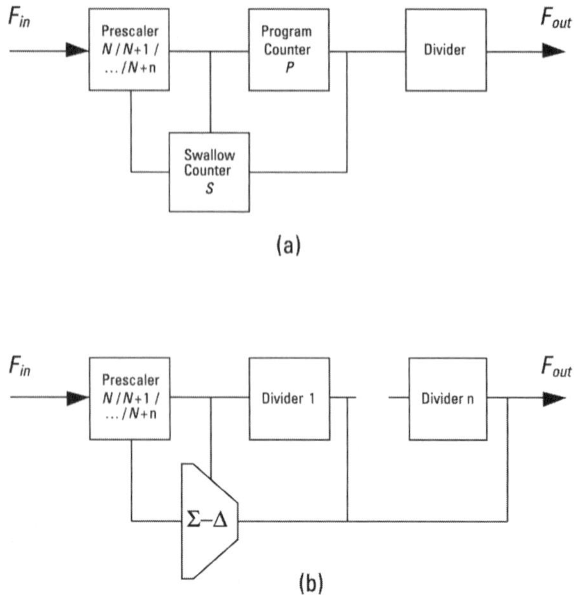

Figure 4.3 Tunable frequency dividers: (a) pulse-swallow structure and (b) sigma-delta structure.

the division ratio, limited to integer numbers in the first block, can only change as a combination of these integers. Whereas in the second case in Figure 4.3(b), the shift between one integer to another in the first divider is not exclusively done at the end of a fixed counter count, producing fractional division ratios, and hence a wider tuning capability.

It is interesting to note that all these circuits are based on the same basic cell, called latch, and for this reason we will dedicate a separate section to the study of this block (Section 4.3).

Focusing our attention on the prescalers, the most commonly used for both types of tunable dividers are the dual modulus prescalers (DMP), which can perform the division by two preselected values. These values can be selected through a specific control input. The different alternatives for this building block are analyzed in Section 4.4.

Once the input frequency has been divided by the DMP, the rest of the blocks usually operate at significantly lower frequencies. For this reason the program counter, swallow counter, and subsequent dividers may be designed using digital libraries. Section 4.5 deals with these blocks.

Finally, it is interesting to note that in many high frequency PLLs the first block of the frequency divider is a divider-by-2. With this strategy the specifications of the rest of the building blocks are conveniently relaxed. The resulting architecture for the PLL would then be the one shown in Figure 4.4.

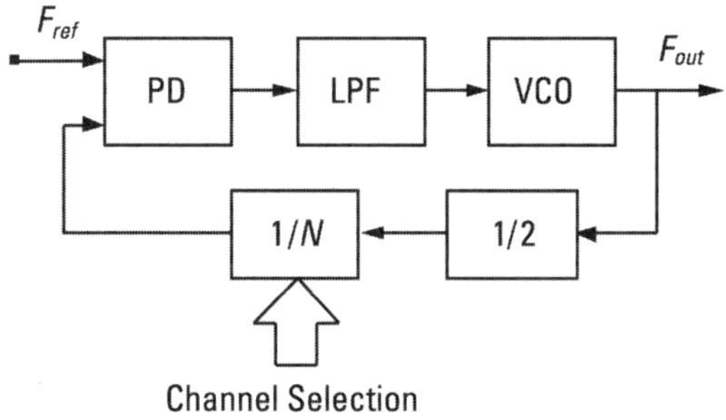

Figure 4.4 PLL with a divide-by-2 circuit as the first module of the frequency divider.

4.3 High-Frequency Divider-by-2

The circuits of division by two, also known as divide-by-two circuits (DTCs), are widely used to generate quadrature outputs. However, because of the fact that these circuits can reach speeds higher than dividers with other factors of division, the DTCs can also be employed in a PLL in order to reduce the VCO frequency down to an appropriate value. This ability means the block is the popular choice in these kinds of systems.

The different architectures found to design DTCs at frequencies in the order of gigahertz can be classified into static latches in master/slave configuration, dynamic latches, and Miller divider latches. Outlined below is the main characteristic of each:

Static Latches in Master/Slave

The first architecture, known also as Johnson counter, is illustrated in Figure 4.5. It consists of two latches connected in master/slave configuration. Each flip-flop is triggered by two complementary clock signals, CLK and $\overline{CLK}$. The two flip-flops work periodically and alternatively between two modes. When the input signal CLK is at a low level, one of the latches is in sensing mode (it

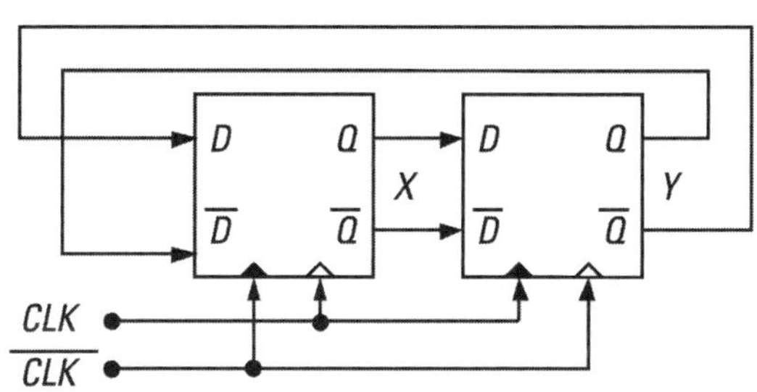

Figure 4.5 Johnson counter block diagram with DTCs.

receives the data at its input *D* and it copies them into its output *Q*), while the other is in latch mode (it keeps the previous output). When the input signal *CLK* goes to high level, the flip-flops exchange their modes of operation. This mechanism means that the frequency from the output signal (*Q* and $\overline{Q}$) can be half that of the input (*CLK* and $\overline{CLK}$).

This architecture provides perfect quadrature in *X* and *Y* only if the signals *CLK* and $\overline{CLK}$ are exactly complementary (i.e., the mismatch between both latches is practically nonexistent). In real conditions the deviations can be translated in differences in phase of up to 5 degrees. Furthermore, if *CLK* and $\overline{CLK}$ are not exactly differential, additional imbalances of phase will appear.

Diverse topologies have been found for the implementation of each one of the latches in this architecture, optimized to achieve maximum operating frequency and minimum power consumption, being the most commonly used the configurations proposed by Razavi, Wang, and the source couple logic (SCL) type.

Dynamic Latches

Unlike statics, the implementation of dynamic latches is not based on a bistable circuit. The parasitic capacity between the nodes acts as a storage element. The most common dynamic architecture employs a true single-phase clocking (TSPC) latch.

Miller Divider

This high-speed method for dividing by two, originally proposed by Miller and explained in generic books (i.e., [3]), is illustrated in Figure 4.6.

Using a mixer and a lowpass filter, this configuration operates in the following way. After carrying out the mixing process, the output from mixer components at frequencies of $F_{in} + F_{out}$ and $F_{in} - F_{out}$ are generated. If the first is suppressed by the filter but the last one is not, then $F_{in} - F_{out} = F_{out}$, and therefore $F_{out} = F_{in}/2$. The simplicity of the loop of transmission allows this topology to operate at speeds higher than half of the f_T of the transistors, regularly achieving the highest speeds out of all divider configurations. However, the phase noise introduced is too high for practical RF applications, since neither the mixer nor the LPF are low-noise-oriented. For this reason, this configuration is not commonly used in high-frequency PLLs.

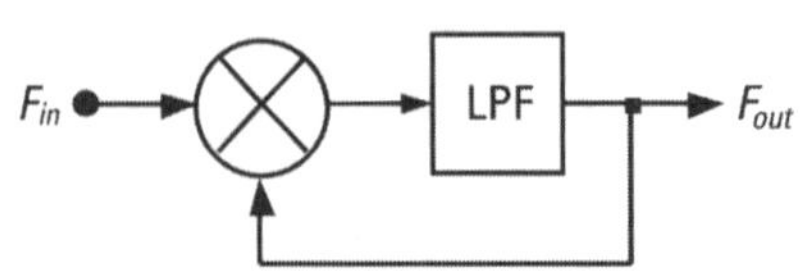

Figure 4.6 Miller divider.

In the next sections, the most common configurations for RF applications are explained in greater detail.

4.3.1 Razavi

The schematic circuit of this latch in master/slave configuration is shown in Figure 4.7 [4].

In high-speed dividers with master/slave topology the slave latch is usually designed as the master dual so that both can be driven by the same clock signal. However, the duality requires one of the latches to incorporate PMOS transistors in the path of the signal, reducing the maximum speed of the divider. In order to avoid this, identical latches that are driven by the complementary clocks CLK and $\overline{CLK}$ are used.

In Figure 4.7 each latch consists of two transistors working in sense mode ($M1$ and $M2$ in the master and $M7$ and $M8$ in the slave), a regenerative loop ($M3$ and $M4$ in the master and $M9$ and $M10$ in the slave) and two transistors acting in pull-up mode ($M5$ and $M6$ in the master and, $M11$ and $M12$ in the slave). When the clock is at high level, $M5$ and $M6$ are cut off and the master

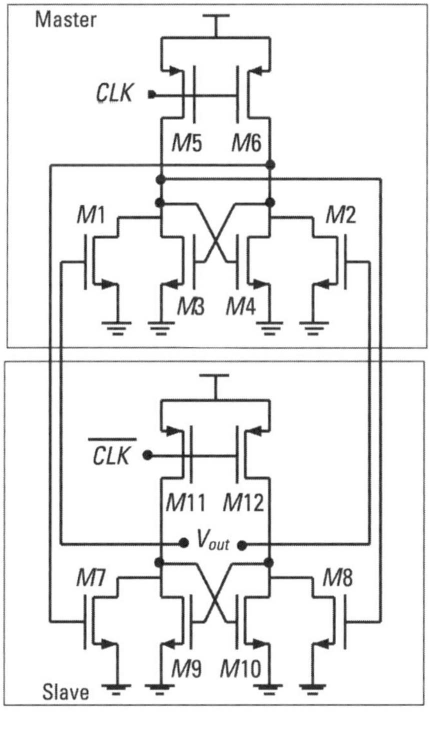

Figure 4.7 Razavi topology for the DTC.

is in sensing state, while $M11$ and $M12$ are activated and the slave is in latch mode. When the clock changes to low level, the opposite process occurs.

The most important advantage of this topology is its speed, given that it does not use stacked PMOS transistors and the current flows just through two gates per cycle. On the other hand, it presents the disadvantage of consuming static energy (it consumes during the whole cycle) and requires a differential clock input signal [5].

4.3.2 Wang

A variation of the Razavi topology has been proposed by Wang [6], whose diagram is presented in Figure 4.8. The differences in this configuration with respect to that of Razavi is the inclusion of NMOS clock transistors in each

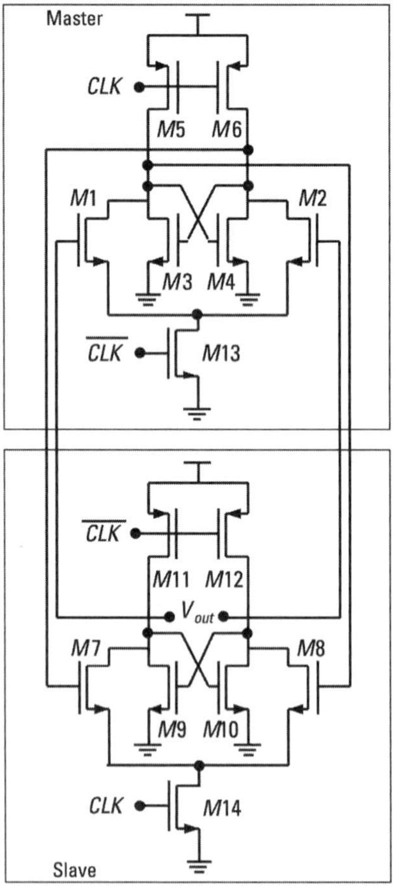

Figure 4.8 Wang topology for the DTC.

one of the two latches (*M*13 in the master and *M*14 in the slave) and the different layout of the PMOS clock transistors.

Despite the similarities between both topologies, the working principle of them is slightly different. In the Wang architecture the operation mode of a flip-flop (sensing or latching) is controlled by a couple of switches *M*13 and *M*14 rather than by the PMOS transistors. Furthermore, the resistance from the PMOS transistors is low in the sensing mode and high in the latching mode, something which makes the time constant smaller and the signal that attacks the other flip-flop higher, in contrast to the Razavi topology.

With this solution, the circuit can operate at higher frequencies, without increasing the power consumption. For example, comparing [4] with [6], the maximum operating frequency of the Wang architecture is superior to 18 GHz with a supply voltage of 1.8V and CMOS technology 0.25 μm. On the contrary, the Razavi topology reaches a working frequency of around 9 GHz for the same supply voltage and a CMOS technology 0.1 μm. On the other hand, the consumption of power in the last one is higher for the same supply voltage. For example, if the supply voltage is 1.8V, the Razavi topology consumes around 10 mW, while that of Wang is 3.6 mW.

4.3.3 SCL

The last topology found to implement each one of the latches in master/slave configuration is the source couple logic (SCL), in which each of the flip-flops are present in the circuital diagram in Figure 4.9 [7].

This topology eliminates the PMOS clock transistors and uses resistances as load. Due to the total absence of PMOS transistors and to the fact that the signal only circulates through two gates per cycle this structure is faster than the two previous ones. In addition, it functions correctly for dynamic ranges of the input clock smaller than in the previous designs [5]. Last, this topology is the most suitable to work at frequencies of various gigahertz. A careful selection of transistor size and type allows for a reasonable equilibrium between speed and power consumption at operating frequencies of gigahertz [2], and provides robustness in the phase noise coming from ground and supply tracks.

On the other hand, it also presents some disadvantages compared to the previous two architectures. As well as the static power consumption and the need for differential clock signals, the occupied silicon area is higher and requires the implementation of current sources [5].

In [7], the implementation of a DTC with SCL topology in a CMOS 0.12 μm technology is described. For a supply voltage of 1.5V an operating frequency of 27 GHz is reached, with a power consumption of 45 mW. It can be seen how the power consumption is superior to that of the Wang and Razavi

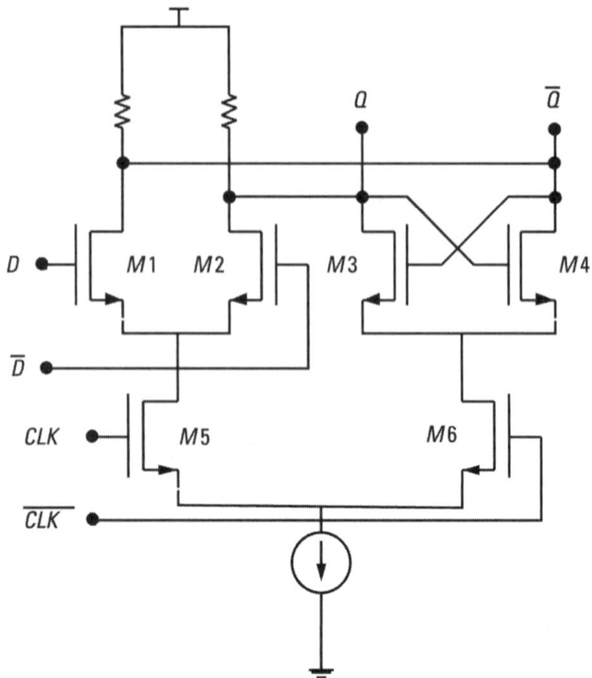

Figure 4.9 SCL flip-flop topology.

topologies for similar supply voltages. However, the operating frequency is also very superior in this SCL topology.

4.3.4 TSPC

The circuital diagram of this dynamic configuration is illustrated in Figure 4.10 [8].

Unlike the static architectures, this topology does not consume static power, dissipating power only in the clock transitions. In addition to this advantage, it can be implemented with less chip area consumption and it does not require differential clock signals. On the contrary, it is a slower topology than the static ones due fundamentally to the use of stacked PMOS transistors and that the signal circulates through three gates per cycle [5].

A deeper theoretical analysis of this DTC topology can be found in [9].

4.4 Dual-Modulus Prescaler

As already explained in the introduction of this chapter, most tunable dividers incorporate a dual-modulus prescaler. This type of high-frequency divider divides

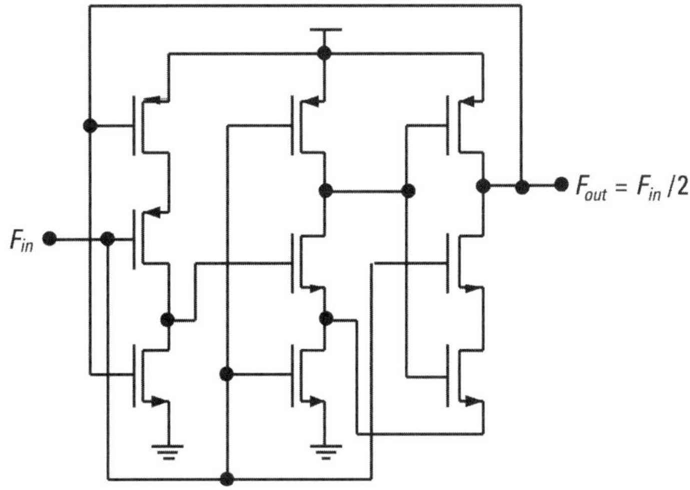

Figure 4.10 SCL TSPC dynamic latch.

the input signal by a rate of division M or $M + 1$ depending on an input control. The values of the division ratio can be very different depending on the requirements of each application.

The prescalers consist mainly of flip-flops and high operating frequency logic gates. The configurations used for implementing high-frequency latches described in the previous section can be used to implement the flip-flops in this type of prescaler.

As there is no need to present a classification of dual-modulus prescalers depending on their division ratio, the following describes the most basic and commonly used type, the prescaler that divides by 2 or 3, whose circuit diagram is illustrated in Figure 4.11(b).

In order to explain the working principle of this circuit, in the following paragraphs we are going to describe the 2/3 divider. Prior to this description, the truth table of a divider by 3 is presented in Table 4.1, while Figure 4.11(a) shows the block diagram. This divider employs two type D flip-flops in master/slave configuration with a logic gate AND in order to create only the three states shown in Table 4.1.

To convert the topology in Figure 4.11(a) into a 2/3 prescaler introducing a logic gate OR between the first flip-flop and the gate AND is sufficient. In this way, when MC is at a high level, the prescaler will divide by 2 and when it is at a low level it will divide by 3. Due to the existence of logic gates between the two flip-flops, this type of divider is slower than the high-frequency dividers by two presented in the previous section.

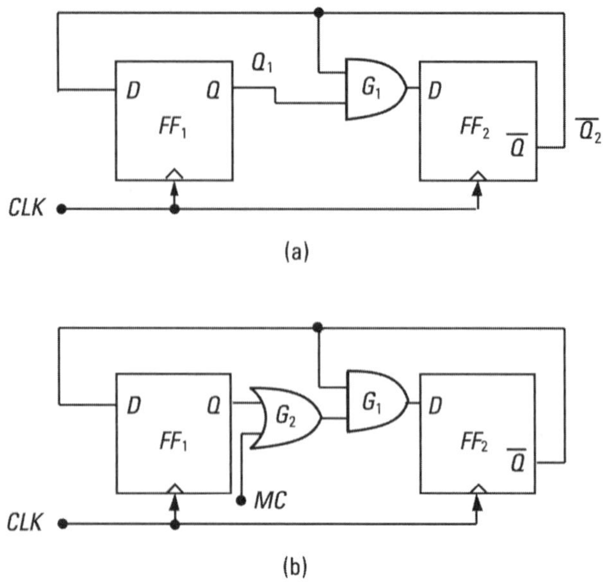

Figure 4.11 (a) Divider by 3 and (b) divider by 2/3.

Table 4.1
Logic State for the Divider by 3 of Figure 4.11(a)

Q_1	0	1	1
$\overline{Q_2}$	1	0	1

4.5 Low-Frequency Dividers

Once the input frequency of the divider has been reduced by using the prescaler, lower-frequency dividers can be employed to complete the division. These dividers are usually called counters. In the case of the pulse-swallow counter topology, the variable divider includes two types of counters: program counter and swallow counter [1], while in the fractional-N architecture a sigma-delta modulator is found.

The program counter is a fixed divider that divides the output frequency of the prescaler (Figure 4.3(a)) by a division rate P. As with the high-frequency dividers it consists of flip-flops and logic gates, but with a fundamental difference: Due to the fact that the frequency in these blocks is often well below the gigahertz range, digital integrated circuit techniques can be used to implement them. An example of a program counter can be seen in Figure 4.12(a), whose

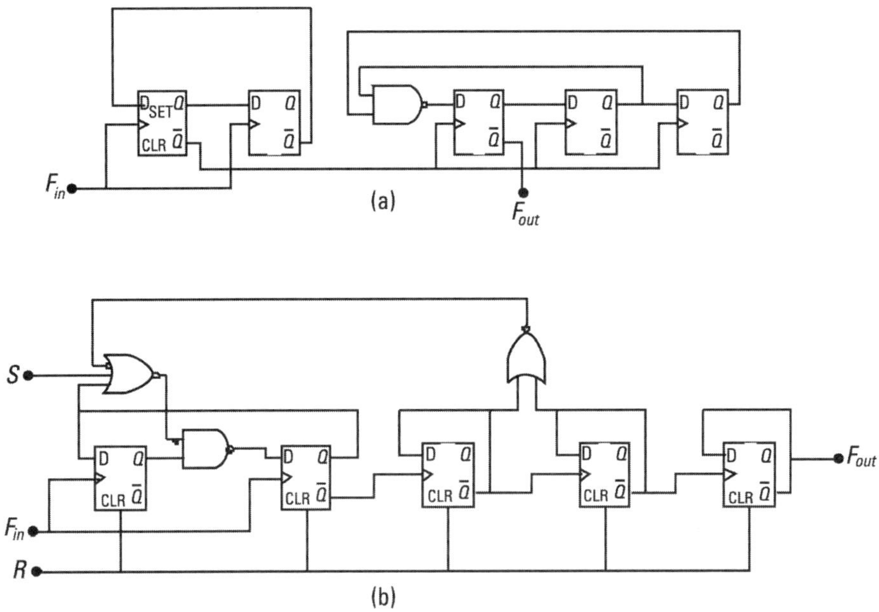

Figure 4.12 (a) Program counter (ratio = 20) and (b) swallow counter (ratio 15/16).

division rate is 20. This divider has been formed using the cascade union of a divider by 4 and one by 5.

The swallow counter is a programmable counter and is used in order to obtain a variable and externally controlled division rate. The input signal is the output of the prescaler and is generally controlled by an external input and by a reset coming from the program counter (Figure 4.12(a)). Its output consists of a line of control that changes the rate of division of the dual-modulus prescaler. The working method of this counter is the following: a certain number is introduced in the counter through the external input control, which starts to count input pulses until this number is reached. At this moment the counter changes the state of its output line so that the prescaler, which was dividing by $M + 1$, goes on to divide by M. Once this point has been passed, the swallow counter carries on dividing its input signal until the program counter counts P pulses and activates the line of reset so that everything returns to its original state. The differences with the high-frequency dividers described earlier apply also to this type of counter. Figure 4.12(b) describes a swallow counter whose rates of division are 15 and 16. Using the external pin S, one of the rates of division is selected, while with the pin R resetting the divider and starting to count the pulses, the original state is achieved.

4.6 Phase Noise

The design of dividers for RF frequency synthesizers is extremely critical given that it requires a tight compromise between high-frequency operation, reduced power consumption and low phase noise. It is known that the influence of phase noise from the divider in the total phase noise at the output of the synthesizer is only relevant below the loop bandwidth. In many cases, it is negligible compared to the contribution of the reference crystal and phase detector. Furthermore, this influence depends on the value of the rate of division (N), given that the phase noise from the divider inside the PLL is multiplied by N^2.

Very few works that analyze the different noise sources of the high-frequency dividers and their influence on the total phase noise have been reported. There are two interesting references, the work proposed by Egan [10], which is rather qualitative, and the more specific reported by [2], which allows for an estimation of the phase noise of an integer variable divider of the pulse-swallow type, employing the SCL architecture to implement the prescaler latches. This last case is especially relevant in order to understand the noise behavior in these blocks.

The theory of noise developed in [2] is applied to a dual-modulus prescaler of a 32/33 rate in CMOS 0.35 μm technology and operating above 3 GHz. In order to compare the model of noise it has been applied to two identical dividers, one with a resynchronization flip-flop at the output and the other without it.

According to this model, the prescaler is the most critical block for the following reasons: (1) it operates at frequencies of several gigahertz, (2) its power consumption is the highest of the synthesizer together with the VCO, and (3) its phase noise is transferred at the output of the locked loop amplified by the square of N.

The prescaler presented in this model has been implemented with a structure that is a combination between a synchronous and asynchronous divider, as can be seen in Figure 4.13(a). Each divider by two consists of two latches in master/slave configuration with SCL architecture and the structure of the 2/3 divider as illustrated in Figure 4.13(b).

The phase noise of the prescaler can affect the total noise at the PLL output. In a divider the simplest way of describing phase noise is in terms of jitter in the time domain. If the output of a flip-flop with superimposed noise is considered, the relation between the power of jitter in a period of the signal and the power of noise in voltage $\left(\sigma_V^2\right)$ can be obtained using (4.1), where S_L is the gradient of the edges of commutation. In this formula, factor 2 takes into account the fact that a period is affected by the noise from the two edges.

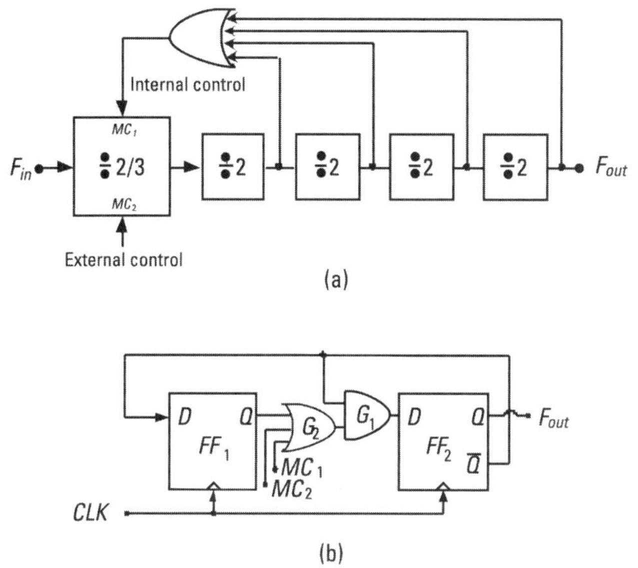

Figure 4.13 (a) Example of a dual modulus prescaler and (b) divider by 2/3.

$$\sigma_{\Delta T}^2 = \frac{2 \cdot \sigma_V^2}{S_L^2} \tag{4.1}$$

For a series of dividers in cascade, the power of jitter at the output is obtained from the quadratic composition of jitter introduced by each divider. The cross-zero jitter of one stage is transferred to the output of the next divider.

As mentioned previously, at the start of this section, the configuration used to implement each flip-flop from the prescaler is an SCL architecture, as described in Section 4.3, but substituting the resistance for PMOS transistors that operate in triode region. It is important to note that the load transistors act as resistances, given that in this region the relation between current and voltage is linear.

The cross-zero jitter can be evaluated using the simplified diagram in Figure 4.14. The sources of noise that can affect this jitter are: thermal noise from the PMOS charge transistors, noise from the current source I_B and noise from the NMOS switches from the inferior differential stage. C_T is the total output capacitance.

The two PMOS load transistors introduce a white noise of the value of $8 \cdot k \cdot T \cdot R$ in the differential output, which is integrated in the band $B \cong 1/(4RC_T)$. As the cross-zero gradient is $S_L = I_B / C_T$, substituting in (4.2), the following value for the power of jitter in a period of the signal is obtained from the PMOS load transistors.

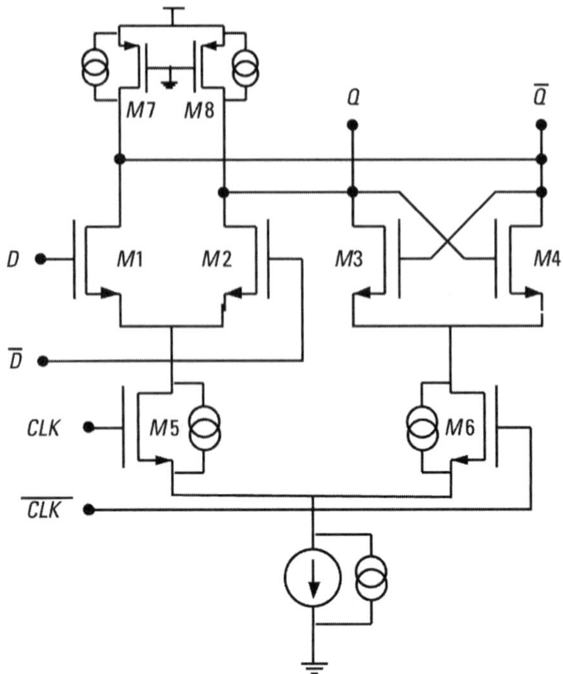

Figure 4.14 Noise source model of the SCL latch.

$$\sigma^2_{\Delta T, PMOS} = 2 \cdot \frac{8 \cdot k \cdot T \cdot R \cdot B}{S_L^2} \cong 2 \cdot \frac{2 \cdot k \cdot T \cdot C_T}{I_B^2} \tag{4.2}$$

Reference [2] shows how the contribution of the other sources of noise in the total jitter is negligible, considering only the noise introduced by the PMOS load transistors.

Once it has been stated that the relevant sources of noise in the latch are the PMOS transistors, it is possible to calculate the total jitter from the prescaler in a period of the signal. Knowing the capacity of the total output from each one of the dividers by two and of the divider by 2/3 and its corresponding polarization currents, the power of total jitter at the output from the prescaler can be estimated using (4.3). As the dividers are in cascade configuration, the total jitter power at the output will be the quadratic composition of jitter introduced by each one of the dividers; that is:

$$\sigma^2_{\Delta T, Total} = 4\sigma^2_{\Delta T, Divisor\_2} + \sigma^2_{\Delta T, Divisor\_2/3} \tag{4.3}$$

The accumulation of the jitter is the worst disadvantage of the asynchronous dividers and can be attenuated through the use of a synchronization flip-flop at the end of the chain. In this case, the total jitter from the output is that generated only by this flip-flop (Figure 4.15). In the particular case of the example presented in this section, the synchronization flip-flop operates at the input frequency of the prescaler (3 GHz), which is why its latches must be laid out as those of the 2/3 high-frequency divider.

Once the total jitter at the output of the prescaler has been calculated, the relation between this jitter and the output phase noise ($L(f_m)$) can be obtained using (4.4), where F_o is output frequency from the divider.

$$\sigma_{\Delta T}^2 = \int_0^{F_o/2} L(f_m) \cdot \frac{\sin^2\left(\frac{\pi \cdot f_m}{F_o}\right)}{F_o^2 \cdot \pi^2} \cdot df_m \tag{4.4}$$

4.7 Layout Considerations

As has already been described in the presentation of the layout of the oscillator in Chapter 3, and with the same objective, several techniques can be been used in order to minimize substrate couplings:

- Ground guard rings to isolate each element;
- Ground contacts in the substrate throughout the circuit;
- Open ground ring and another supply ring around the divider;
- Decoupling capacitors between the supply voltage path and ground.

In addition, the criteria followed in the design of the metal tracks of interconnection of the integrated oscillator, with respect to the maximum current that they can support and parasitic effects that they introduce, should be also applied to the high frequency divider.

The part of the layout of this component that requires the most precise implementation is obviously the one corresponding to the higher-frequency blocks. As has already been seen in the design of the circuital diagram of the

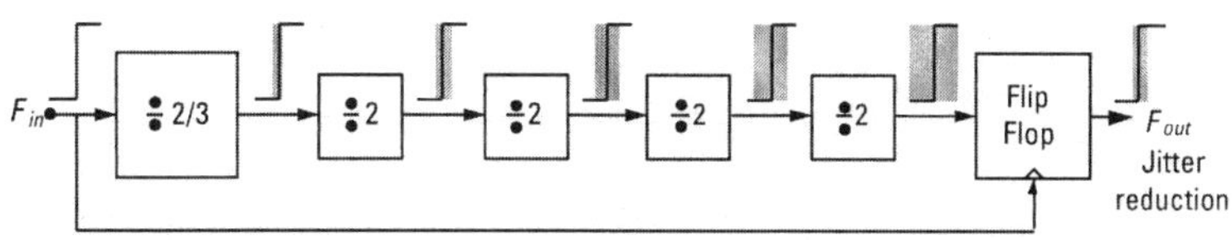

Figure 4.15 Effect of the synchronization flip-flop.

device, the jitter and consequently the phase noise introduced by each high-frequency flip-flop is directly proportional to the total capacitance resulting from the sum of the output node's capacitance plus that of the input node of the next stage. Therefore, the interconnection tracks in this stage must be made as short as possible and in the most external metal layer (which is usually the thickest) with the intention of minimizing the introduced parasitic capacitance. Furthermore, given that this metal possesses low resistivity, the parasitic resistance introduced will also be very small. In addition, in order to minimize the parasitic capacitance between the metal tracks in the high frequency stage, it is desirable to avoid track crosses.

A more detailed explanation of the design process of each block is included in Chapter 8.

References

[1] Egan, W. F., *Frequency Synthesis by Phase Lock*, New York: Wiley-Interscience, 2000.

[2] Levantino, S., et al., "Phase Noise in Digital Frequency Dividers," *IEEE Journal of Solid-State Circuits*, Vol. 39, No. 5, May 2004, pp. 775–783.

[3] Razavi, B., *RF Microelectronics*, Upper Saddle River, NJ: Prentice-Hall, 1998.

[4] Razavi, B., K. F. Lee, and R. H. Yan, "Design of High-Speed, Low-Power Frequency Dividers and Phase-Locked Loops in Deep Submicron CMOS," *IEEE Journal of Solid-State Circuits*, Vol. 30, No. 2, February 1995, pp. 101–109.

[5] Perrott, M., "High Speed Communication Circuits and Systems: High Speed Frequency Dividers," Mitopencourseware, Massachusetts Institute of Technology, 2003.

[6] Wang, H., "A 1.8V 3mW 16.8GHz Frequency Divider in 0.25 μm CMOS," *IEEE International Solid-State Circuits Conference*, 2000, pp. 196–197.

[7] Wohlmuth, H., and D. Kehrer, "A High Sensitivity Static 2:1 Frequency Divider up to 27 GHz in 120nm CMOS," *Proc. Eur. Solid State Circuits Conf. (ESSCIRC),* 2002, pp. 823–826.

[8] Yuan, J., and C. Svensson, "High-Speed CMOS Circuits Technique," *IEEE Journal of Solid-State Circuits*, Vol. 24, No. 1, February 1989, pp. 62–70.

[9] Navarro Soares, J., and W. A. M. Van Noije, "A 1.6-GHz Dual Modulus Prescaler Using the Extended True-Single-Phase-Clock CMOS Circuit Technique (E-TSPC)," *IEEE Journal of Solid-State Circuits*, Vol. 34, No. 1, January 1999, p. 97.

[10] Egan, W. F., "Modeling Phase Noise in Frequency Dividers," *IEEE Trans. on Ultrasonics, Ferroelectrics, and Frequency Control*, Vol. 37, No. 4, July 1990, pp. 307–315.

5

Phase Frequency Detector/Phase Detector

A phase frequency detector/phase detector generates an output signal proportional to the phase and/or frequency difference that exists between its two input signals. One of the input signals is fixed, with very stable frequency and generally generated by a quartz crystal. The other input signal can be variable, less stable and comes from the output of the oscillator after passing through the frequency divider. The function of the phase frequency detector/phase detector within the loop is to correct the excess of phase that exists between two inputs and to lock the frequency by means of a slight variation in the VCO voltage control.

Currently four main types of phase frequency detector/phase detectors are used. The first category is analogical or multiplying phase detectors that are based on the multiplication of two sinusoidal signals from the same frequency. The second and third categories are sequential circuits that operate with the information provided by the zero crossings of the input signals. Among these we can highlight the second and the third categories for the logic gate OR-exclusive and the flip-flop-based phase detector, respectively. The fourth category is a sequential circuit that also provides a signal dependent to the frequency: this is a PFD, in comparison to the three first types which are only PD. These blocks are aimed to lock the loop when the PLL output signal is unlocked.

In this chapter each one of the previous configurations are presented with an explanation of their main advantages and disadvantages. In Section 5.5 the different approximations found for the calculation of phase noise are explained. The chapter ends with design recommendations.

5.1 Multipliers

If the two phase detector inputs are sinusoidal, a *mixer* or multiplier can be used as a phase detector. In order to clarify this point, let us consider two signals as shown in (5.1) and (5.2). If we use these signals as inputs of a balanced *mixer* phase detector, the output generated signal would follow (5.3).

$$v_1 = A_1 \cdot \sin(\omega_1 \cdot t + \theta_1) \tag{5.1}$$

$$v_2 = A_2 \cdot \cos(\omega_2 \cdot t + \theta_2) \tag{5.2}$$

$$v_d = A_d \cdot \{\sin[(\omega_1 - \omega_2) \cdot t + \theta_1 - \theta_2] + \sin[(\omega_1 + \omega_2) \cdot t + \theta_1 + \theta_2]\} \tag{5.3}$$

where the amplitude of the signal resulting from the product (A_d) is given in (5.4).

$$A_d = \frac{A_1 \cdot A_2}{2} \tag{5.4}$$

In phase-locked loop condition, the two angular frequencies are the same and the dc component at the phase detector output equals $A_d \cdot sen(\theta_1 - \theta_2)$, which is proportional to the phase difference for small values of $(\theta_1 - \theta_2)$.

This phase detector is especially useful in applications where the reference frequency is too high for other solutions. The main disadvantage of this type of phase detector is the high number of undesired frequency components generated at the output, which have to be attenuated by the loop filter. The most disturbing is the component situated at the frequency sum $(\omega_1 + \omega_2)$, which in locked condition will be at twice the reference frequency, and consequently is directly added to the first harmonic spurious peak of the reference tone.

5.2 Exclusive-OR Logic Gate

An exclusive-OR (EXOR) logic gate can be used as a phase detector, as demonstrated in Figure 5.1. This figure presents the functioning of an EXOR when two slightly out of phase signals *A* and *B* arrive at each one of their inputs.

The two input signals *A* and *B* are shown in Figure 5.1(b), together with the output signal *C*. Its relative mean value is proportional to the phase difference on a half-cycle range, as shown in Figure 5.1(c). The operating point of this phase detector must be chosen in the middle of the linear range of its transfer

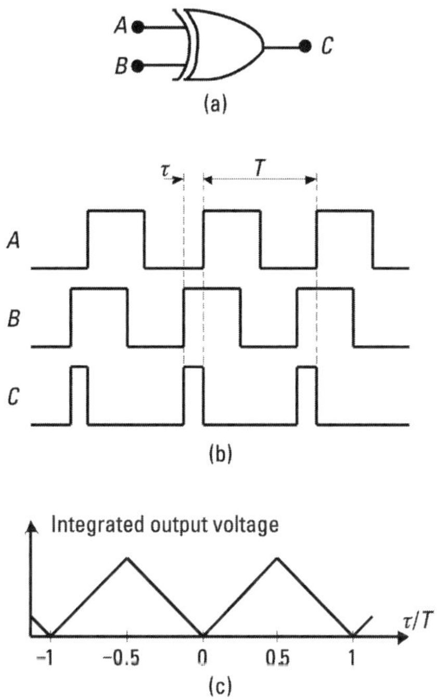

Figure 5.1 (a–c) OR-exclusive phase detector.

function shown in Figure 5.1(c), which corresponds to a phase difference of 90° between its inputs. From this operating point the phase difference between *A* and *B* is allowed to shift 90° in both directions, achieving a maximum working range of 180°, with, then, no impact on the duty cycle of A/B.

5.3 Flip-Flop

A *set-reset* (S-R) *flip-flop* can also be used as a phase detector as illustrated in Figure 5.2. A narrow pulse train in both inputs *A* and *B* activate and deactivate output *C*. The average value of *C* presents a sawtooth waveform, with a linear range of a complete cycle (360°).

5.4 PFD/CP

The phase-frequency detector with charge pump (PFD/CP) is a sequential phase detector with the ability of also locking the frequency that is mostly used for the implementation of locked loops because:

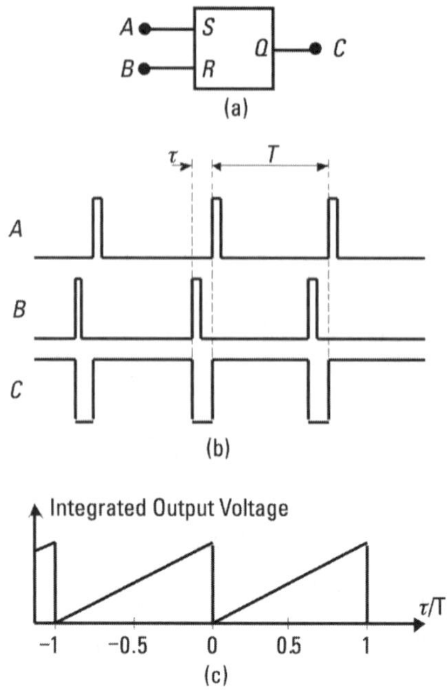

Figure 5.2 (a–c) Flip-flop phase detector.

- It presents an input linear range of ±360°.
- It acts as a phase detector while the loop remains locked and provides a signal proportional to the difference in frequency that contributes to lock the PLL.
- The circuit schematic is compatible with CMOS integrated circuit technologies.

Figure 5.3 shows a block diagram of this type of detector.
As the name indicates, the circuit consists of two distinct parts:

1. The phase-frequency detector (PFD), responsible for generating voltage pulses of a width proportional to the phase and frequency difference between its inputs;
2. The charge pump (CP), responsible for delivering a charge proportional to this phase/frequency difference.

A simple implementation of the PFD, using type D flip-flops, is shown in Figure 5.3. Before we discuss the complete system, it is useful to remember the operation of D flip-flops: when the clock input is at zero it stays invariable,

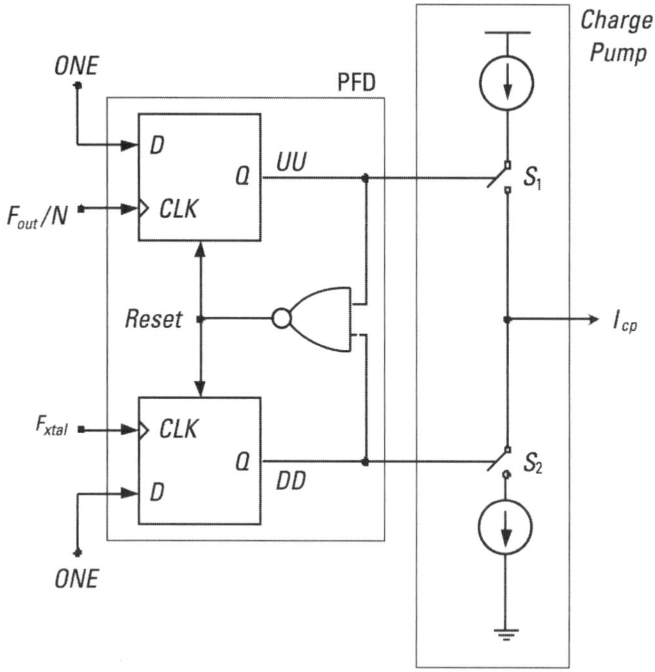

Figure 5.3 Phase frequency detector with charge pump (PFD/CP).

and when the clock input changes at a high level the data input is copied to the output. Independently of the clock and the data input, when the reset input is activated the output of the flip-flop goes to a low level.

The PFD consists of two D flip-flops and a NAND gate. As an initial condition, it is assumed that both outputs (*UU* and *DD*) are reset (at a low level). When the first edge arrives by either of the two clock inputs, the flip-flop in question will copy the input data (always connected to a high level) into its output. On the other hand, the other flip-flop will continue as at the beginning (with the output at a low level) until the first positive clock edge arrives. When this happens it will copy the input data *D* (also at a high level) to its output. At this point the two outputs of the flip-flops happen to be at a high level, something which will cause the logic gate AND (or NAND) to activate its output and reset both flip-flops. From this point everything starts again from the beginning.

Figure 5.4 shows this explanation in a graphic way. It is important to point out that during the time that the NAND gate takes to reset both flip-flops a small pulse at the output of one of the flip-flops will appear (in the flip-flop in which the rising edge arrives later), which has not been represented due to its short duration (of the order of few nanoseconds).

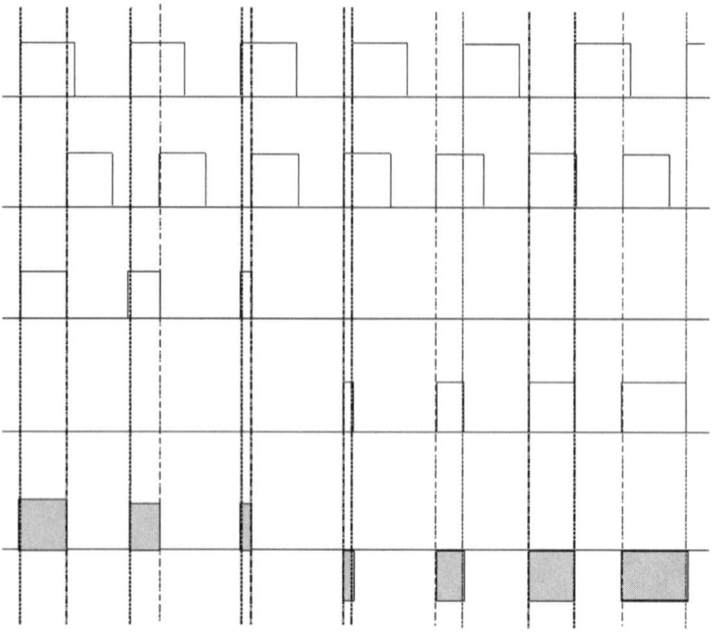

Figure 5.4 Functioning of the PFD/CP phase detector of Figure 5.3.

The working principle of the CP can be described as follows: if the line *UU* is active, the charge pump generates positive current pulses (acting as source); on the contrary, if *DD* is active these current pulses will be negative (drain). In normal operation, one of the inputs of the PFD is connected to the reference crystal (F_{xtal}), while the other comes from the oscillator output after being divided by the module N (F_{out}/N) divider. This process is illustrated in Figure 5.4.

The output current drives the lowpass filter of the loop, which is responsible for transforming the pulse train into a voltage value.

Figure 5.5 is a representation of the mean current from the output of the detector (I_{PD}) for a small frequency error and slow phase modifications versus the phase difference between the two PFD inputs, where I_{cp} is the module of the output current of the charge pump.

This type of phase detector, despite being the most used in the implementation of locked loops [1–5], presents nonideal effects that may degrade the PLL overall performance, as a consequence of a modulation in the VCO line of control. The most important of these problems is dead zone.

The dead zone of a PFD/CP phase detector is the phase difference between its two inputs for which the charge pump does not inject current in the loop filter. Figure 5.6 shows a more realistic output current than the ideal depicted

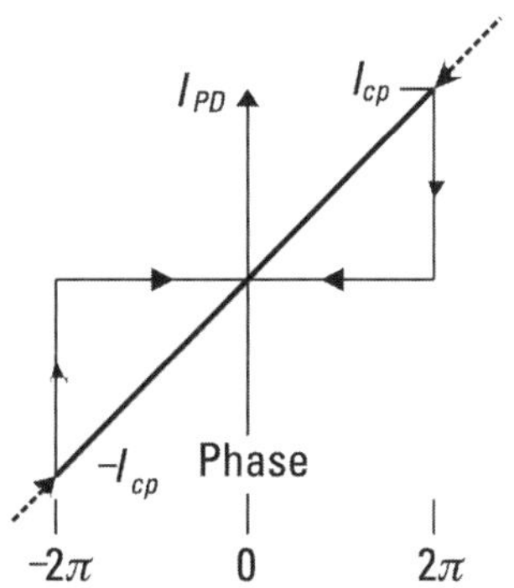

Figure 5.5 PFD/CP phase detector output characteristic.

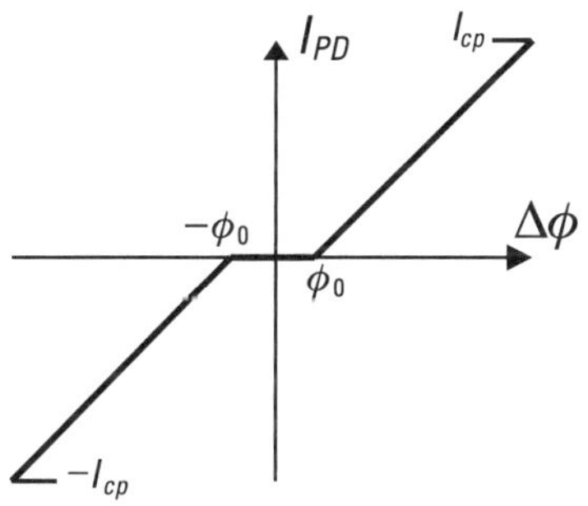

Figure 5.6 Dead zone of the PFD/CP phase detector.

in Figure 5.5. Here, the output current for phase differences less than ϕ_0 is almost zero. This dead zone provokes the rising of the jitter at the output of the loop; that is, an increase in phase noise of the PLL.

One of the most used methods to eliminate the dead zone consists of including an even number of inverters in the reset of the PFD flip-flops that do not change its logic state, but give rise to a delay in the signal large enough to cancel it. In this way, in ideal locked loop situation with null phase difference, the two outputs of the PFD generate narrow voltage pulses of the same width that activate the charge pump and inject current pulses of the same amplitude in the loop filter. As a consequence of this, the dead zone is eliminated.

But while the dead zone is eliminated, the reset pulses also introduce negative effects. The most important are due to the delay that exists between the pulses from *UU* and *DD* (Figure 5.3), the difference in absolute value between the charge and discharge currents from the charge pump and the finite capacitance seen from the current source drain of the charge pump.

As the charge pump needs both to supply and drain current pulses, one of the two switches of the charge pump must be implemented with PMOS transistors. Due to this, it is necessary to introduce an inverter at the corresponding PFD output to ensure the correct commutation. This gate creates a small delay in one of the two commutation paths, which gives rise to two small current pulses of the same amplitude and width, but with different sign and phase between each other. This does not modify the VCO voltage control because the total charge injected is zero, but it provokes a periodic jitter in this voltage, creating undesired reference spurs.

The second negative effect comes from the difference in absolute value between the charge and discharge currents from the charge pump. If this occurs, in each moment of the comparison the pulses from the PFD output in locked condition give rise to a positive or negative charge injection that displace the VCO voltage control by a determined value. In order to solve this, the PLL introduces a phase error between the input and the output in a way that the total current injected by the charge pump in each cycle is zero. In this way, one of the current pulses will have a bigger amplitude but will be narrower. It is useful to mention three important issues related to this effect: (1) the VCO voltage control will also experience a periodic jitter, (2) the mismatching between the charge and discharge currents depends on the voltage of the output from the charge pump, and (3) the mismatching of current injection between the two transistors that act like switches in the charge pump and the feedthrough effect from the clock also increase the phase error and the jitter in the VCO tuning voltage. The second effect is also another fundamental cause of reference spurs.

The third and last effect is known by the name of *charge sharing*. For a better understanding of this problem Figure 5.7 shows the implementation of a simple charge pump, where the transistor switches are represented in a symbolic way by S_1 and S_2, and the current sources are created by transistors ($M1$ and $M2$) controlled by an input voltage. The parasitic capacitances C_X and C_Y seen

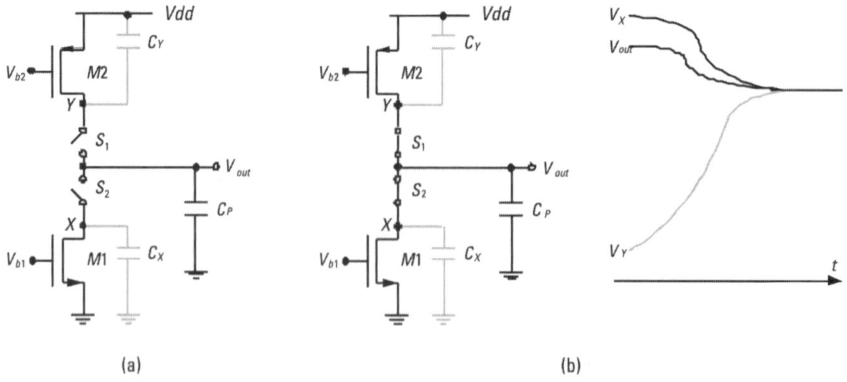

Figure 5.7 (a, b) Charge sharing effect between C_P and the capacities in X and Y.

from the current source drains contribute notably to the generation of this effect.

In the first place it is assumed that the switches in Figure 5.7 are found in a deactivated state, allowing for $M1$ to discharge X to ground and $M2$ to charge Y to V_{dd}. In the following instant of phase comparison, S_1 and S_2 are activated, V_X increase and V_Y decreases, in a way that both voltages reach a value approximately the same as V_{out} neglecting the voltage drop through S_1 and S_2 (Figure 5.7(b)). But after the switches have been activated, including if the phase error is zero, the charge and discharge currents are exactly the same and $C_X = C_Y$, the voltage of the output of the V_{out} cannot remain constant. For example, if V_{out} is relatively high, V_X a bigger quantity than V_Y will vary. The difference between these two variations must be supplied by C_P, which induces a small variation in the voltage of the V_{out} output.

5.5 Phase Noise of Phase Detectors

The contribution of the phase detector to total phase noise from the PLL, is considerable and makes itself apparent within the bandwidth of the loop. Generally, as discussed in earlier chapters, at offset frequencies with respect to the carrier superior to the bandwidth of the PLL, the dominant noise comes from the VCO, while at inferior frequencies the phase detector and the reference crystal usually dominate.

Among the reported works about phase noise calculation, there are two interesting studies that offer some valuable help to understand the problem. The first, proposed by Craninckx [6], is based on the current noise generated by the charge pump during the period of time in which it is active. The second, proposed by Banerjee [7], estimates this factor of merit from the level of noise of the phase detector normalized to the reference frequency of 1 Hz ($L_{1\text{Hz}}$). This parameter is notably experimental and difficult to predict precisely. Both noise-oriented models have been analyzed theoretically in Chapter 2.

5.5.1 The Craninckx Model

This model estimates the noise from the phase detector based on the noise of the current from the charge pump (I_{CP}) during the time it is active. In locked loop condition, if both current sources have a magnitude of I_{cp} and both are active during a fraction of time α_{cp}, the charge pump can be represented by a current noise source whose value is given in (5.5).

$$\frac{\overline{i^2}}{\Delta f} = 2 \cdot \alpha_{cp} \cdot 4 \cdot k \cdot T \cdot \frac{2 \cdot I_{cp}}{(V_{GS} - V_t)_{cp}} \tag{5.5}$$

where $(V_{GS} - V_t)_{cp}$ is the difference between the gate-source voltage and the threshold voltage of the transistors of the charge pump current sources.

Equation (5.5) allows for the calculation of phase detector noise as current noise from a charge pump but in an isolated way. What is really interesting is to calculate this noise at the output of the loop. The contribution of the detector at the PLL output is considerable at offset frequencies less that the bandwidth of the loop. In addition, it can be estimated using (5.6), which is only valid for offset frequencies much less than the bandwidth of the PLL.

$$L\{\Delta\omega \ll \omega_c\} = \frac{1}{2} \cdot \left(2 \cdot \pi \cdot \frac{N}{I_{cp}}\right)^2 \cdot 2 \cdot \alpha_{cp} \cdot 4 \cdot k \cdot T \cdot \frac{2 \cdot I_{cp}}{(V_{GS} - V_t)_{cp}} \tag{5.6}$$

Therefore, the phase noise from the charge pump can be reduced by increasing the value $(V_{GS} - V_t)_{cp}$ of the transistors from the current sources or by decreasing the fraction of time during which the charge pump is active. Also a correct choice of I_{cp} is important in order to obtain a good characteristic of phase noise.

5.5.2 The Banerjee Model

This model estimates the phase detector noise at the output of the loop using the parameter $L_{1\text{Hz}}$, which represents the level of noise from the PLL within its bandwidth ($\omega \ll \omega_c$) for a reference frequency of 1 Hz. In addition to the phase detector, other sources of noise exist within the bandwidth of the loop (VCO, crystal, and so forth). However, it is usually this last source of noise which is the most significant, something that makes attributing this parameter almost totally to phase noise possible.

In [7], typical values of phase detector noise normalized at 1 Hz are presented for different commercial synthesizers that vary between –199 and –211 dBc/Hz.

This parameter allows for comparison of PLLs that work at different reference frequencies, but presents the disadvantage of being an eminently empirical parameter, making it of little use from the point of view of the designer.

From $L_{1\text{Hz}}$ the noise of the phase detector at the output of the loop in dBc/Hz within the bandwidth of the PLL can be obtained easily using (5.7).

$$L\{\Delta\omega \ll \omega_c\} = L_{1\text{Hz}} + 10 \cdot \log(F_{ref}) + 20 \cdot \log(N) \tag{5.7}$$

where N is the ratio of division of the PLL and F_{ref} its reference frequency.

5.6 Practical Considerations of Design

The objective of this section is to indicate the basic aspects that need to be considered for the implementation of the phase detector, from design to its schematic level and to the complete development of its layout.

5.6.1 Design of the PFD

There are three fundamental steps in the design of the PFD:

1. Estimation of the Delay of the Reset Path

   The most important aspect of the design of this block is without doubt the estimation of the delay of the reset path. As mentioned earlier, one of the main disadvantages of the PFD/CP topology is the dead zone that appears in its transfer characteristic.

   In order to minimize the dead zone it is necessary to increase the width of the pulses until they reach a valid logical level and so are able to activate the charge pump. In order to achieve this it is necessary to introduce an appropriate delay in the reset path.

   The minimum delay that must be introduced in the reset path has been estimated as the average value of the time it takes the pulses in the output nodes *UU* and *DD* to rise and fall [8]. This delay can be introduced by including inverters. Logically, the number of gates will always have to be even in order to maintain the reset state of the flip-flops.

2. Scaling of the Logic Gates

   Once the calculation of the number of necessary inverters in the reset path has been carried out, the following decision that needs to be taken is the scaling of the logic gates the PFD is comprised of.

   When dealing with a component that operates digitally and at a frequency not normally very high, the same criteria can be applied as in the case of the low-frequency digital divider. On the other hand, selecting a bigger channel width for the PMOS transistors with respect to that of the NMOS is recommended so that they compensate for the different characteristics of both types of transistors.

3. Delay Block

   As far as a delay is necessary for the elimination of existing phase difference between two signals in the nodes *UU* and *DD* (Figure 5.3), the channel length and width of the transistors of this block must be selected in a way that the delay is cancelled in the middle of the rising edge of the pulses from both signals.

5.6.2 Design of the Charge Pump

The design of this second block of the phase detector is the most influential on the level of the reference spurs at the output of the loop, which is why it is necessary to take into account certain considerations to try to reduce it. An example of the charge pump scheme is illustrated in Figure 5.8.

In general terms, the charge pump is composed of a commutation section, formed in this case by *M*1, *M*2, *M*3, and *M*4 transistors, and another section allocated to replicating the current.

The critical size corresponds to the transistors that intervene in the commutation of a simple switch like the one in Figure 5.9, consisting of an NMOS

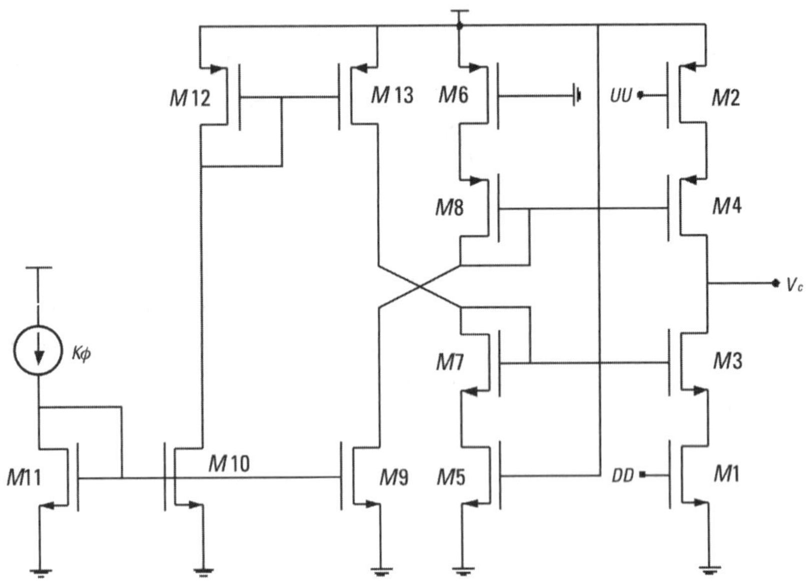

Figure 5.8 Circuital scheme of the PFD charge pump.

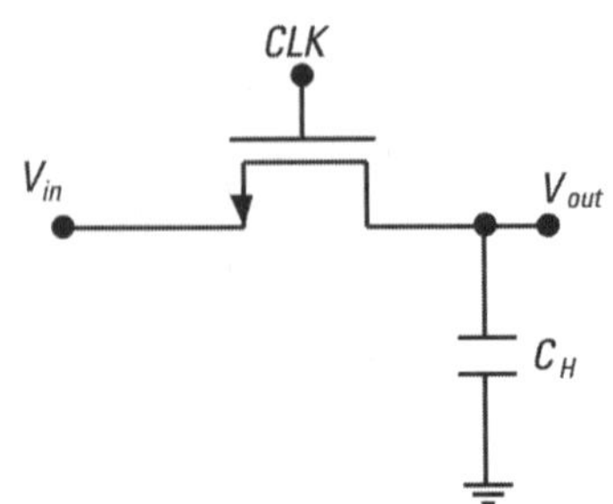

Figure 5.9 Circuital scheme of a basic switch.

transistor plus a sampling capacitor C_H, depending principally on the channel resistance of this transistor that operates in triode region and on the C_H capacitance. To increase this speed it is advisable to reduce the channel resistance, something that implies increasing the W/L relation of the transistor and decreasing the sampling capacitance.

In addition to depending on the W/L ratio, the channel resistance is inversely proportional to the mobility of the electrons (or gaps) in the channel, which is bigger in an NMOS transistor than in that of a PMOS one (usually three times bigger). In this way, the NMOS transistor, with the same dimensions, presents a lower channel resistance and as a consequence a higher commutation speed.

In Figure 5.8 it can be observed that the transistors that act as switches ($M1$ and $M2$) swing between cutoff and triode regions, while $M3$ and $M4$ oscillate between cutoff and saturation. In this way, when each one of the two branches of the charge pump is driven the $M3$ and $M4$ transistors add another resistance (output resistance in saturation) in series with that of conductance (resistance in triode) from $M1$ and $M2$. This saturation resistance is also inversely proportional to the W/L ratio.

The triode (r_{triode}) and saturation (r_{sat}) resistances mentioned earlier can be estimated using (5.8) and (5.9), respectively [9].

$$r_{triode} = \frac{1}{\left[\mu \cdot C_{ox} \cdot \frac{W}{L} \cdot (V_{GS} - V_t)\right]} \tag{5.8}$$

$$r_{sat} = \frac{1}{\left[\frac{\lambda \cdot \mu \cdot C_{ox} \cdot W}{2 \cdot L} \cdot (V_{GS} - V_t)^2\right]} \tag{5.9}$$

where μ is the mobility of electrons or holes in the channel, C_{ox} is the capacitance per unit of area of the parallel capacitor formed between the gate and channel, V_{GS} is the voltage difference between the gate and the source of the transistor, V_t is the threshold voltage of these, and λ is the parameter that takes into account the phenomenon known as channel-length modulation.

From all these considerations, the procedure proposed for the scaling of the transistors is:

1. Choose identical dimensions for the $M1$ and $M2$ transistors so that they are capable of commuting the operating frequency of the charge pump correctly.
2. Select the size of $M3$ and $M4$ so that the speeds of commutation of both branches of the pump are the same. In this way, as the NMOS

transistors are faster than the PMOS, the W/L relation of $M4$ needs to be superior to that of $M3$ in order to reduce the total conduction resistance from the superior branch and increase the speed of commutation.

3. Take into account the fact that the parasitic capacitance introduced depends on the dimensions of the selected transistors. As mentioned earlier, the speed of commutation decreases accordingly with the increase in the speed of sampling C_H. In this way, a situation can arise in which by trying to increase the speed of commutation using an increase of W/L, this effect is compensated by an increase in C_H up to the point where no improvement in speed can be noticed.
4. To obtain a more precise scaling it is frequently necessary to use a simulator and to establish the next objective: If there is no phase difference between the PFD inputs, the charge and discharge currents generated by the two branches should be exactly the same. In this way, in locked loop condition the charge and discharge currents will be practically the same and the total current injected into the filter will decrease notably. With this strategy the jitter in the VCO voltage control is reduced, and also the level of the reference spurs. Logically, in addition to the speed of commutation, the synchronization of $M1$ and $M2$ is also important, something which has already been improved by introducing the delay element in the PFD.

5.7 Design of the Layout of the Phase Detector

Due to the fact that the operating frequency of this block is usually much lower than in the rest, the introduced parasitic effects have a lesser effect. The most important are those introduced in the charge pump's branches of commutation, as has been explained previously which influence the speed of commutation of the switches. This is why the interconnection tracks of the elements if these two branches (NMOS and PMOS) must be as short as possible.

The following are the most important considerations that need to be taken into account at the time of carrying out the layout of the PFD:

- Maintaining the symmetry between the layout of the flip-flops is recommended, as between the logic gates belonging to each of the branches of the PFD. As described earlier, the synchronization between the charge pump input signals from the charge pump is fundamental in order to reduce the level of reference spurs at the output of the PLL.

- It is advisable that the two branches of the phase detector have the same length (including the delay element) so that the dephase between the PFD output signals (nodes *UU* and *DD*) is as little as possible.

With regards to the charge pump, in null dephase condition between the PFD inputs (locked loop) the charge and discharge currents from the charge pump must be exactly the same so that the total current injected into the loop filter is as reduced as possible. In order to assure this the following considerations are recommended [10]:

- As the influence of the silicon stress on the mobility of the two carriers depends on the orientation along the crystal axis of reference, the groups of transistors in which reducing the errors of matching is desired will have to be orientated along the same axis of reference. The groups of transistors from the charge pump in which matching techniques have been applied are those that present identical dimensions and are also of the same type (NMOS or PMOS).
- As MOS transistors are vulnerable to temperature gradients, stress, and thickness of the oxide, the distance between them must be reduced as much as possible in order to minimize these effects.
- Implementing the layout of the transistors for which reducing the matching errors is desired as compactly as possible. This means making a suitable choice of the number of digits for each transistor.
- Wherever is possible, apply common centroid configurations. In order to use this technique it is necessary to divide each transistor into an even number of digits and place them alternatively.

References

[1] Leung, G. C. T., and H. C. Luong, "A 1-V 5.2-GHz CMOS Synthesizer for WLAN Applications," *IEEE Journal of Solid-State Circuits*, Vol. 39, No. 11, November 2004, pp. 1873–1882.

[2] Herzel, F., G. Fischer, and H. Gustat, "An Integrated CMOS RF Synthesizer for 802.11a Wireless LAN," *IEEE Journal of Solid-State Circuits*, Vol. 38, No. 10, October 2003, pp. 1767–1770.

[3] Rategh, H., H. Samavati, and T. Lee, "A CMOS Frequency Synthesizer with an Injection-Locked Frequency Divider for a 5-GHz Wireless LAN Receiver," *IEEE Journal of Solid-State Circuits*, Vol. 35, No. 5, May 2000, pp. 780–787.

[4] Zhang, P., et al., "A 5-GHz Direct-Conversion CMOS Transceiver," *IEEE Journal of Solid-State Circuits*, Vol. 38, No. 12, December 2003, pp. 2232–2237.

[5] Vassiliou, I., et al., "A Single-Chip Digitally Calibrated 5.15-5.825-GHz 0.18-μm CMOS Transceiver for 802.11a Wireless LAN," *IEEE Journal of Solid-State Circuits*, Vol. 38, No. 12, December 2003, pp. 2221–2229.

[6] Craninckx, J., and M. Steyaert, *Wireless CMOS Frequency Synthesizer Design*, Norwell, MA: Kluwer Academic Publishers, 1998.

[7] Banerjee, D., *PLL Performance, Simulation, and Design*, 2nd ed., National Semiconductor, 2001.

[8] Lee, H., et al., "Scheme for No Dead Zone, Fast PFD Design," *Journal of the Korean Physical Society*, Vol. 40, No. 4, April 2002, pp. 543–545.

[9] Sedra, A. S., and K. C. Smith, *Microelectronic Circuits*, 4th ed., New York: Oxford University Press, 1998.

[10] Hastings, A., *The Art of Analog Layout*, Upper Saddle River, NJ: Prentice-Hall, 2001.

6

Determination of Building Blocks Specifications

One of the most crucial design steps is the determination of realistic building block specifications. Decisions reached at this point will drive the rest of the design process, and therefore have a strong impact on the time and effort spent. On the other hand, the methodology needed to carry out this task is difficult to standardize. There are many different factors that must be integrated and considered, for example:

- The working characteristics of the device under consideration;
- The limitations of the communication standard;
- The initial requirements that must be observed, some of them may be imposed by commercial requirements (cost, mass-production) or company strategy (technology, software);
- The previous experience of the design team is a key element in determining the limits of the different building blocks;
- The available information, which can bring some light to the initial design stages.

In this chapter, we propose a methodology to determine the initial building block specifications of the PLL. The flow is explained with a practical example, a PLL for the 5-GHz U-NII band, as we believe it is the best way to understand the procedure. In Section 6.1 we comment on the initial requirements found at the beginning of the work. Section 6.2 shows the architecture selected for the PLL. Following this, Section 6.3 is dedicated to the exposition of the specific software tools developed for this case. Finally, Section 6.4 shows the study of each building block.

6.1 Initial Requirements

The first step of the design procedure is the identification of the initial requirements. Obviously, these requirements are driven by the application in which the PLL is required. In this case we are going to assume three initial conditions:

- Output frequency: 3.2 GHz;
- Fabrication technology: standard CMOS 0.18 μm;
- Application: WLAN receiver for the 5-GHz U-NII band.

These are the determining conditions that are imposed by the general project, in this case, the development of a fully integrated CMOS WLAN receiver for the U-NII 5-GHz band. The front-end has been implemented following a heterodyne low-IF architecture, as displayed in Figure 6.1.

In this figure BPF1 and BPF2 are the passband filters from the receiver, LNA the low noise amplifier, MIX1 and MIX2 the frequency mixers, LO1 and LO2 the local oscillators, CAG1 and CAG2 the controllable gain amplifiers, and LPF1 and LPF2 the low-pass filters for the end of the chain. Specifically, the frequency synthesizer used as an example in this book corresponds to the second frequency conversion (LO2), whose operating frequency has a value of 3.2 GHz. As it is apparent from the system, the synthesizer under study (PLL2) has fixed output frequency, since the channel selection is performed in PLL1.

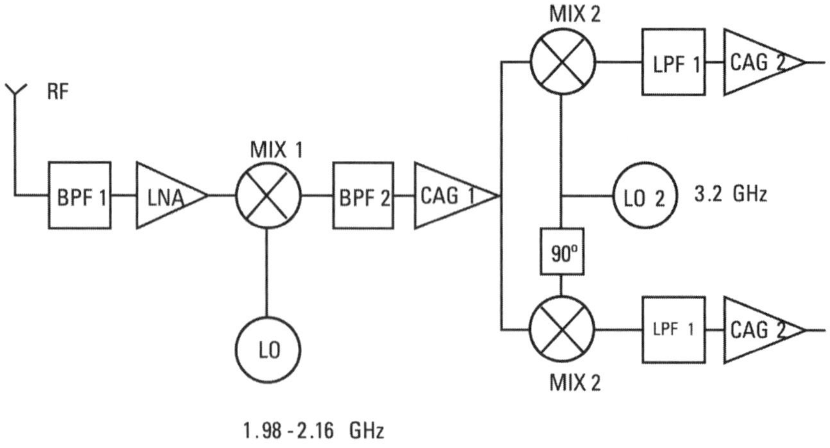

Figure 6.1 Heterodyne architecture for the receiver for the WLAN IEEE 801.11a standard.

With this information, we can extract more relevant input data for the design of the PLL, for example we can define the reference crystal, the phase noise requirements, the maximum spurious emissions, and the lock time of the device. In the following paragraphs we outline the values selected for this example.

6.1.1 Previous Works Search

The first step in any design realization is the revision of previously reported works. It is always interesting to check the limitations and problems found by other researchers. In our case, we searched information related to OFDM integrated receivers, RFIC working in the 5-GHz band and CMOS implementations. Table 6.1 shows the most relevant papers selected.

Obviously, the number of published works is much higher than the list presented in this table. These works have been selected for the clarity of the exposition and they held a general interest regarding our design.

6.1.2 Reference Crystal

The two requirements of the reference crystal determined in this section are the output frequency and its maximum acceptable tolerance in parts per million (ppm).

Table 6.1
Previously Reported Works Used as References

Frequency (GHz)	Description	Application	Technology	Reference
5.2	Transceiver	802.11a/b/g	CMOS	[1]
5.2	Transceiver	802.11a/b/g	CMOS	[2]
5.2	Receiver	802.11a	0.18 CMOS	[3]
5.2	Receiver	802.11a	0.18 CMOS	[4]
5.3	Progr. divider	Hiperlan	0.25 CMOS	[5]
1.3	VCO	—	0.35 CMOS	[6]
1.8	PLL	—	0.35 CMOS	[7]
1.9	VCO	—	0.5 CMOS	[8]
1.1	VCO	GSM	0.8 CMOS	[9]
4.3	VCO	—	0.35 BiCMOS	[10]
2	VCO	—	0.65 BiCMOS	[11]
5.05	VCO, mixer	WLAN	0.8 SiGe	[12]
2.5	VCO	—	Bipolar	[13]
5.2	VCO, LNA, mixer	WLAN	0.25 CMOS	[14]
5	VCO, LNA, mixer	WLAN	0.24 CMOS	[15]
2.2	LNA, mixer	WLAN	0.6 CMOS	[16]
1.8	LNA, mixer	—	0.8 SiGe	[17]

The reference frequency is obviously dependent on the architecture of the synthesizer that is going to be used. As it is shown in Section 6.2, the architecture proposed for this work is the locked loop with a fixed-N frequency divider. Since the output frequency of the PLL under study is fixed, there are many valid values for this reference frequency. However, it is a good practice to select a value equal to the channel spacing, therefore avoiding potential in-channel interferers produced by harmonics of this reference or undesired couplings through the PCB or connection tracks. In our case, WLAN channels are 20 MHz wide, and consequently, this is the value selected for the reference frequency.

It is interesting to note that this decision affects also the loop bandwidth. If the reference frequency is excessively small, the loop bandwidth will be even smaller and as a consequence the lock time of the PLL will be too high.

Regarding the accuracy in ppm, this requirement has been obtained from the standard IEEE 802.11a [18], which establishes a maximum reference crystal tolerance of ±20 ppm.

6.1.3 Phase Noise

The appropriate method for the determination of phase noise requirement is strongly dependent on the targeted application. In some cases the standard completely determines the phase noise mask, while in others, as it is our situation with IEEE 802.11a, there may be less information provided.

The phase noise mask can be estimated with two requirements:

1. Phase noise for an offset frequency greater than the loop bandwidth;
2. Specification of the value of the integral of total phase noise from the PLL over the channel bandwidth.

Figure 6.2 shows the graphical representation of these two concepts, which completes the detached analysis of the phase noise performed in Figure 2.6.

This is a simplified approach that is based on the approximation that for frequencies greater than the loop bandwidth, the noise is dominated by the VCO contribution, and therefore the slope of the curve is –20 dB/dec. For frequencies closer to the loop bandwidth the situation is more complicated, because of the effect of the rest PLL building blocks, primarily the PD/PFD. But in this case the integral of the phase noise over the channel provides more information to define the mask.

The complete definition of PLL noise mask needs more inputs, some of them not defined at this point of the design flow, including the contribution of PLL building blocks or the loop bandwidth. As all these concepts present cross-influences and trade-offs, it is recommended to follow an iterative process

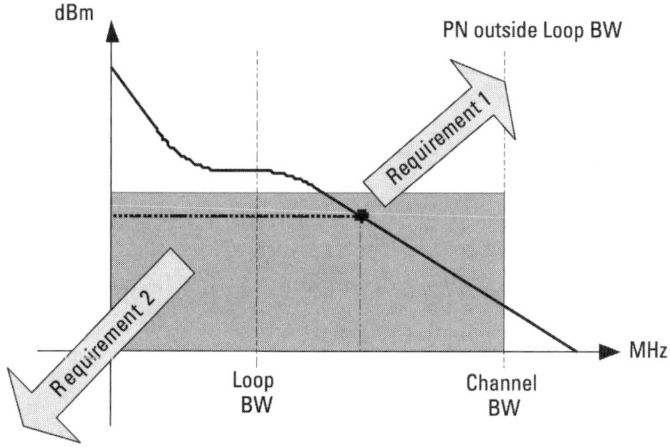

Figure 6.2 Phase noise mask.

for the selection of the final values. In this work, a simulation tool has been implemented to facilitate these calculations, as shown in Section 6.3.

In the following sections, the determination of the two main requirements for phase noise calculation is presented.

6.1.3.1 PLL Phase Noise for Offset Frequency Greater Than the Loop Bandwidth

The noise can be produced by two phenomena, mixing with adjacent channel interferences and modulation phase impairments, which in a basic explanation are the real effects due to the mixing among the subcarriers of the modulated channel. In most of the cases, the first effect can be neglected in comparison with the closer subcarrier separation taken into account for the second effect.

In order to determine the specification for the noise provoked by modulation phase impairments, let us focus first on the consequences of this effect. The phase noise in the OFDM systems introduces intercarrier interference (ICI) and leads to degradation (D) in the SNR [19]. Equation (6.1) provides this relation.

$$D = \frac{11}{6 \ln 10} 4\pi\beta T \frac{E_s}{N_0} \tag{6.1}$$

where β is the single-sided –3-dB linewidth of phase noise power spectral density (assumed to have a Lorentzian spectrum), $1/T$ is the subcarrier spacing (312.5 kHz for IEEE802.11.a), E_s is the symbol energy, and N_0 is the single-sided noise PSD. E_s/N_0 is a relation depending on the *SNR*, *DR*, the data rate and BW_{eff} standard communication bandwidth (6.2).

$$\frac{E_s}{N_0} = SNR - 10\log\left(\frac{DR}{BW_{eff}}\right) \tag{6.2}$$

Equation (6.3) provides the expression to calculate the phase noise (*PN*) at an offset frequency (f_{off}) depending on β, which can be calculated using (6.1).

$$PN(f) = 10\log\left(\frac{1}{\pi}\frac{\beta}{f_{off}^2 + \beta^2}\right) \tag{6.3}$$

At this point of the design process, the loop bandwidth is not defined yet. However, we can select a value for a suitable out-of-loop frequency (f_{off}) if we assume the rule of thumb that loop bandwidth is usually 100 to 10 times smaller than channel bandwidth. In our case the latter is 20 MHz, therefore we can select the value of 1 MHz as a good approximation for the offset frequency.

In order to complete the calculation, let us assume a 0.1-dB degradation in the *SNR* and with a 64-QAM modulation, which according to the IEEE 802.11a has DR = 54 Mb/s and BW_{eff} = 16.25 MHz and SNR = 26. Then we follow the next steps:

- Equation (6.2) gives E_s/N_0 = 20.85 (nondimensional ratio);
- At this point, (6.1) allows the calculation of β, resulting 149.8 s^{-1};
- Finally, with (6.3) we can obtain the phase noise value, in this case PN = −103.2 dBc/Hz at 1 MHz.

This value is consistent with the recommendation found on previously reported works. For example, [20] provides a more conservative specification of −110 dBc/Hz at 1 MHz.

6.1.3.2 Specification of the Value of the Integral of Total Phase Noise from the PLL over the Channel Bandwidth

According to [21], the integral of phase noise from a frequency synthesizer for a WLAN-OFDM system at 5.25 GHz over the bandwidth of the channel must be less than or equal to 32 dBc in order to achieve acceptable values of the BER and of the *signal-to-noise ratio* (SNR) in relation to the channel. The calculation of this parameter is not straightforward, nevertheless in the referred paper, Côme et al. present the fifth section dedicated to the influence of the phase noise on the BER performance of OFDM transceivers, as one of the main front-end nonidealities. The influence is due to two effects: the first, common to all the subcarriers, is the close-in phase error that rotates the whole constellation, but this can be estimated and corrected. On the other hand, the second effect results in intersubcarrier interference that cannot be corrected and has a higher frequency Gaussian-like noise behavior, as the authors explain.

The integration of the phase noise power density function (PDF) over the channel bandwidth is a good approximation for the BER performance. This PDF is normally modeled by a Lorentzian function with uniform phase distribution. As a result in this paper, the simulation of this function for Gaussian channel with a 64-QAM modulation and a coding rate of 3/4 results in a degradation for the BER lower than 0.5 dB with a total integrated phase noise K of −32 dBc. This has then been the selected value for the example presented in this book.

6.1.4 Spurious Emissions

As described in Section 6.2, the architecture of the frequency synthesizer presented in this book is a locked-loop type synthesizer with an integer-N frequency divider, the only spurious emissions that exist are those that appear at offset frequencies multiple of the reference frequency. Therefore, it is only necessary to determine the specification for this type of emission.

The level of reference spurs is deduced from the interference of the adjacent channel and assuming that the maximum speed of transmission for the standard considered is 54 Mbits/s. For this maximum speed, the minimum sensitivity stated by the standard for the receiver is −65 dBm. On the other hand, the maximum power level at the input of the receiver must be −30 dBm, also in accordance with the standard. By means of these two inputs a dynamic range of 35 dB at the input of the receiver is obtained. Then, considering an error margin of 5 dB, the interference from the adjacent channel can be 40 dB higher than the power of the desired channel. For this interference from the adjacent channel, the requirement of the reference spurs in dBc can be obtained using (6.4).

$$Spur = -40 - SNR = -40 - 19 = -59 \text{ dBc} \tag{6.4}$$

where an SNR relation of 19 dB has been assumed for a BER of 10^{-6} and a modulation 64-QAM [20].

6.1.5 Lock Time

This requirement has been taken directly from the IEEE 802.11a standard, which specifies that the maximum time that can be used in the commutation of a channel to another must be less than 1 ms [18].

Despite the fact that this specification is not strictly necessary for the frequency synthesizer that is used here, given that its output frequency is fixed, its value has been given so that it can be used in future designs and in order to be able to compare it with the results provided for any frequency synthesizer parametric simulator.

6.2 Architecture Selection

As has already been presented in Chapter 2, the phase locked loop is the most common type of synthesizer used in integrated applications for such high-frequency and low-phase noise conditions. The decisions that must be taken at this point are those listed:

- Division ratio of the divider;
- Complexity of the phase detector;
- Order of the passive loop filter.

As stated in this example's introduction, the divider selected here has a fixed integer-N structure, since the channel selection is provided by the first PLL of the receiver. The reference selected is 20 MHz and then the division ratio is 160, to achieve the 3.2-GHz requirement.

The differentiation among the PLL's structures is based on the two other decisions for the phase frequency detector/phase detector and the loop filter which combination and order establishes the loop bandwidth that affects the locking time.

The simplest structure of a PLL is composed of a frequency detector and a RC filter for fixing the control of the VCO. These blocks, combined with the oscillator, create a second-order system, the simplest possible. Nevertheless, a circuit that can detect and act to lock both phase and frequency is proven to improve the acquisition range and consequently the locking time of the loop.

As result, the first improvement that can be added to the basic second-order system is the phase detector, which converts the comparative block of the loop into a phase frequency detector. The output of this block can be converted into a DC signal to control the VCO in two different manners: the low-pass filter used until this point or a charge pump. The second alternative presents a better static phase-locking since there is no decay of the charge between the instants of the phase comparison [22].

Nevertheless, the PFD in association with the CP supposes the addition of two imaginary poles [Im(s) in the denominator of (2.8)] to the system and an extra zero [Ex(s) in the numerator of (2.8)] is then required in the open loop transfer function to stabilize the closed loop. One solution is to place a resistor in series with the capacitor of the initial RC filter. Unfortunately, this can introduce ripple and the locking range can be compromised. But with a second capacitor connected from the output of the charge pump to ground, the system dynamics fit our goal.

Figure 6.3 shows the building block diagram finally selected for this work.

With these assumptions we shape the first tentative scheme. It's important to be aware that this is just a starting point. This system must be now simulated in the context of the complete receiver, and the result of these simulations could provoke the rejection or improvement of this scheme.

6.3 Ad Hoc Simulation Tool: Simusyn

Once the first design assumptions have been taken, the optimization of the PLL system must be carried out with the typical iterative process in which the different building block characteristics are also set out. This is a complex process that requires the use of software tools. The most straightforward solution to this problem is the use of commercial CAD tools for electronic circuit and system simulation. However this strategy can lead to some problems:

- Lack of understanding of the already implemented toolboxes or templates;
- Loss of direction due to the high volume of information available and often not relevant for the design case;
- Difficulties in finding information about the models actually implemented.

In order to keep full control of the calculations we recommend the development of an ad hoc simulation tool, at least to carry out the first simulations. The objective of this tool should be the determination of the most relevant design parameters, and also to gain insight into the influence of the variation of design variables into these parameters. Although this step could be deemed

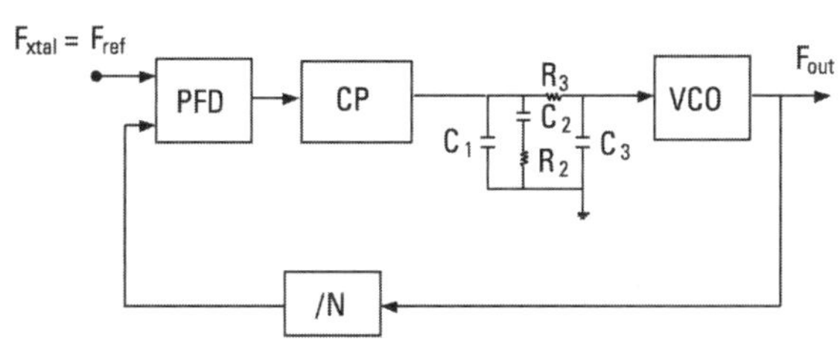

Figure 6.3 Building block diagram of the PLL implemented in this work.

as unnecessary, the use of these tools can speed up the design process by reducing the number of design variables. At this juncture it is important to clarify this point: This statement should not be understood as a negative criticism of commercially available software system design. These tools offer invaluable help for the design of complex systems, when analog and digital signals must be combined or when we try to analyze working under complex conditions. However, when these tools are combined with ad hoc tools for building block design, we believe the designer is in full command of the situation.

In the next sections we present the software we implemented for this project, called Simusyn. The platform we used is MATLAB, but each designer can choose any generic tool for it. The objective of the program is to calculate the phase noise, spurious emissions and the lock time of the PLL based on a series of initial design variables. The routine also determines the components of the loop filter. The program can be downloaded at www.ceit.es/electrocom/RF/dowloads.htm, and a more detailed explanation can be found in [23].

6.3.1 Simusyn Description

Simusyn is a program that allows the quick calculation of phase noise, spurious emissions, and the lock time of a PLL. Its mission is to provide a first set of values for the PLL design, but it can be used also in the most exact determination of PLL's factors of merit once each of its functional blocks has been implemented.

The models upon which Simusyn is based are those proposed by Banerjee, which have been described in previous chapters of this book. The input parameters selected for Simusyn are:

- Ratios of reference and division;
- Frequency of the reference crystal;
- Extra attenuation of the spurs ($ATTEN$);
- VCO gain (K_{VCO});
- Phase margin of the loop (ϕ_p);
- Phase detector noise normalized to the reference frequency of 1 Hz ($L_{1\text{Hz}}$);
- Bandwidth of the loop (BW_{LOOP});
- Output frequency from the PLL;
- Current of the charge pump (K_ϕ);
- Leakage current from the charge pump ($Leakage$);
- Output frequency range of the PLL if necessary ($|f_{\max} - f_{\min}|$);
- Accuracy (tol) for calculating the lock time;

- *BaseLeakageSpur* and *BasePulseSpur* constants;
- Phase noise from the VCO, divider and reference crystal.

This simulation tool also has the option of choosing between a passive filter of the second order, a passive filter of the third order, and an active filter with SGA topology.

In Figure 6.4, the main working window of Simusyn is illustrated. It can be seen that the phase noise is represented in a graphic way, while the values of the loop filter components, the level of spurious emissions, and the lock time are expressed in a numeric way.

The sources of noise that are taken into account in Simusyn are the VCO, the phase detector, the frequency divider, the reference crystal, and the resistance and gain (for active devices) of the loop filter.

6.3.2 Models Implemented in Simusyn

As it has already been described in Section 2.2.1, the phase noise from a synthesizer can be divided into noise inside and outside the bandwidth of the

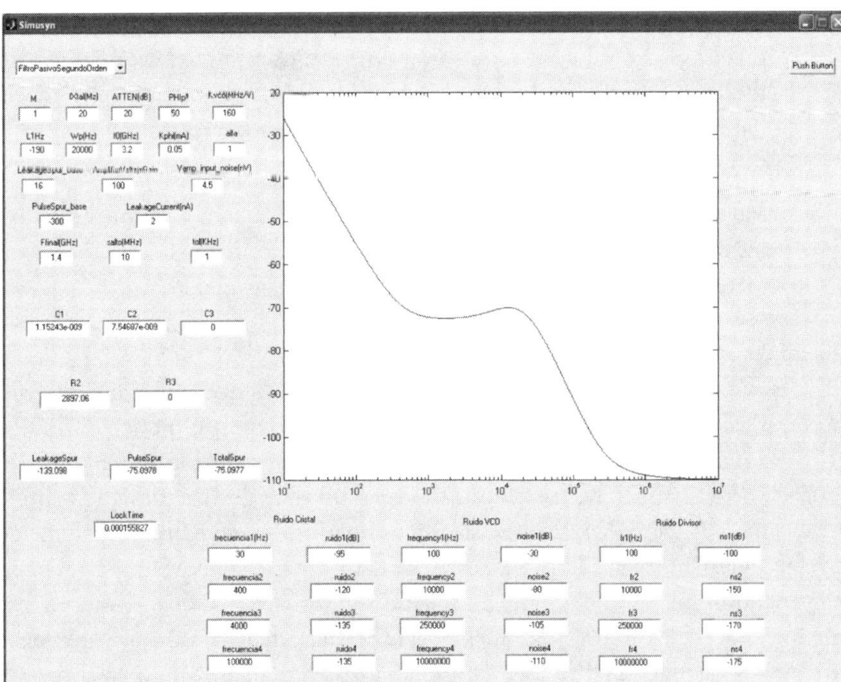

Figure 6.4 Window screen of the design tool Simusyn.

loop. The noise outside of this bandwidth is mostly dominated by the VCO, which is widely analyzed and modeled in Section 3.4. In addition, electrical circuit simulation tools permit accurate estimation of this merit factor, FOM, which means that the prediction of the total phase noise from a synthesizer at offset frequencies superior to the bandwidth of the PLL does not constitute an important design problem.

On the other hand, the phase noise inside the bandwidth is more difficult to predict. It is usually found to be dominated by the crystal and the phase detector, although sometimes it is also necessary to take into account the phase noise from the divider:

- The noise introduced by the reference crystal can be easily calculated with a spectrum analyzer [24] or from the datasheet when available.
- The noise associated with the phase detector requires more attention. As described in Section 5.5, two models have been found in the previously reported works. One of them is rather optimistic and the other depends on $L_{1\mathrm{Hz}}$, an eminently empirical parameter. The solution chosen for Simusym is to use the model put forward by Banerjee to predict this noise, using CAD simulation results to progressively specify with higher precision the value of $L_{1\mathrm{Hz}}$. For this purpose, the iteration starts with the least favorable value of $L_{1\mathrm{Hz}}$ found in the bibliographic references. Later on, this figure is corrected following the results of the VCO simulation.
- Finally, the noise from the divider inside the bandwidth of the loop is initially ignored. This assumption must be confirmed once the results of the actual performance of the divider are available.

As with the phase noise, the estimation of reference spurs also depends on empirical factors. Once again, we have selected Banerjee models to predict this parameter (see Section 2.2.2). As described in this section this model uses empirical constants. For this reason, the method used for the design is similar to the previous case: A rather pessimist estimation is assumed, and then it is corrected with more accurate data provided by circuit simulation.

In addition to the phase noise and spurious emissions, Simusyn also estimates the lock time of a frequency synthesizer when its output experiences a frequency change. The model adopted was presented in Section 2.2.3.

Finally, as well as predicting the three previous aspects, Simusyn also calculates the values of the components of the loop filter.

6.4 Building Block Specification

The most relevant specifications of each one of the functional blocks that constitutes the frequency synthesizer are explained in this section. As a result, the last section shows the list of global loop parameters used as inputs for Simusyn in order to be complete the initial simulation of the system.

6.4.1 Reference Crystal

The two specifications of the reference crystal needed to carry out the initial simulation of the system are the output frequency and its phase noise.

6.4.1.1 Output Frequency

The output frequency from the crystal can be deduced from the requirement for the reference frequency, as established in Section 6.1.1. In this case, the frequency of the crystal coincides with the reference frequency, which is equal to 20 MHz.

6.4.1.2 Phase Noise

The crystal oscillator that has been selected is the MFO-280F from KSS, which has been especially designed to fulfill the IEEE 802.11a, b, and g standards. This oscillator presents a frequency accuracy of ±15 ppm, better than ±20 ppm tolerance required in the IEEE 802.11a specification, and its phase noise can be seen in Figure 6.5.

In Figure 6.5 it can be observed that the phase noise from the reference crystal oscillator is of the order of –94.4 dBc/Hz at an offset frequency of 10 Hz and falls with a 20-dB/decade slope.

6.4.2 VCO

The specifications of the VCO explained in this section are the output frequency, the frequency range, the gain, the amplitude of the output, and the phase noise.

6.4.2.1 Output Frequency

The output frequency used in this example, whose value is 3.2 GHz, has already been determined in Section 6.1.

6.4.2.2 Frequency Range

The frequency range that the VCO must cover in this specific case has been calculated taking into account the tolerances of the manufacturing process of the LC-tank inductor and varactor, therefore adding the corresponding errors to their respective circuit models. If the output frequency was variable, adding this variation to the total calculation of the range would also be necessary.

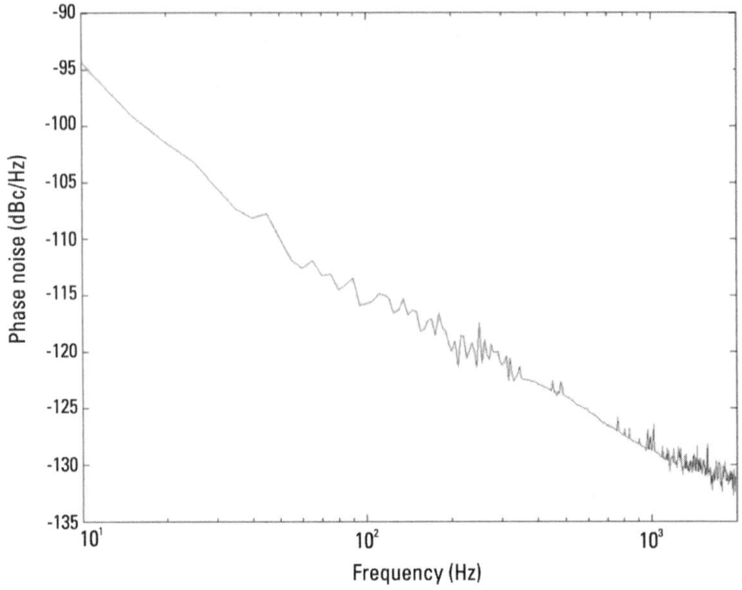

Figure 6.5 Phase noise from the reference crystal.

In order to take this decision, a careful study of fabrication technology must be carried out. In our case, an error of ±10% has been assumed in the circuit models of integrated passive components (capacitances, varactors, and inductors) from the tank. These tolerances are overall technology-dependent. From this value, the superior extremes (f_{HIGH}) and inferior (f_{LOW}) from the frequency range of the VCO can be calculated using (6.5) and (6.6), respectively.

$$f_{HIGH} = \frac{1}{2 \cdot \pi\sqrt{L \cdot C \cdot 0.9}} = 1.054 \cdot F_{out} = 3{,}373 \text{ MHz} \tag{6.5}$$

$$f_{LOW} = \frac{1}{2 \cdot \pi\sqrt{L \cdot C \cdot 1.1}} = 0.953 \cdot F_{out} = 3{,}051 \text{ MHz} \tag{6.6}$$

where F_{out} is the output frequency from the VCO. It is important to point out that from these equations it is found that the VCO needs to cover a frequency range of 322 MHz.

6.4.2.3 VCO Gain

Having calculated the frequency range it is now possible to calculate the VCO gain. Here it is useful to remember that the gain calculated in this section is actually an average value. In fact, the speed of variation of the VCO frequency is not the same for all the VCO voltage controls given that it depends on the variation of the capacity of the tank varactors with regards to this voltage. As the real characteristics of the varactors used have not been defined at this point of the design flow, the quotient between the frequency range calculated previously and the range of the control voltage from the oscillator will be taken as the value of the VCO gain. In this way, considering a voltage control range of between 0.5V to 2.8V in order to avoid significant mismatches in the charge pump, the VCO gain can be obtained through (6.7).

$$K_{VCO} = \frac{f_{\max} - f_{\min}}{Vcontrol_{\max} - Vcontrol_{\min}} = \frac{322}{2.3} = 140 \text{ MHz/V} \qquad (6.7)$$

Nevertheless, this gain value is determined in a more accurate way in the design phase of each one of the PLL functional blocks.

6.4.2.4 Output Amplitude

This specification is fundamentally determined by the voltage that the subsequent stage (a mixer) needs at its input in order to operate at optimum conditions. In the case of this project the minimum voltage, $V_{amplitude}$, which is required at the VCO has been estimated at 0.5V. This value represents a power from the local oscillator signal of –9 dBm applied to a mixer input Z_{in} of 1,000Ω, as calculated from (6.8).

$$P_{\text{dBm}} = 10 \log\left(\frac{\left(\frac{V_{amplitude}}{\sqrt{2}}\right)^2}{Z_{in}} * 1{,}000\right) \qquad (6.8)$$

6.4.2.5 Phase Noise

The selection of the initial value of phase noise has been accomplished following the worst scenario conditions, which corresponds to the requirement determined in Section 6.1.1.1 (–110 dBc/Hz at 1 MHz).

6.4.2.6 Viability of VCO Specifications

Once the first set of values have been determined, it is essential to check that these specifications are within the working limits of the fabrication technology selected. Certainly it is not always possible to ensure that the final implemented circuit will reach the specs, but at least it is useful to analyze the viability of the starting point and throughout this process, we may also identify the challenging points of the design.

The technology selected for the implementation of this PLL is CMOS 0.18 μm, which clearly permits working frequencies superior to 5.35 GHz for active components. As a rule of thumb, its transition frequency, f_T, reaches 50 GHz as described by the main foundries, and then at frequencies lower than five times less (10 GHz in this case), the transistors are still available. Regarding passive components, we must ensure that their quality factor are sufficient to achieve the specification of the phase noise. For this purpose the Craninckx model of phase noise presented in Section 3.4.2.1 has been used, given that this model permits obtaining a mathematical expression for its estimation.

In this way, from (6.9) this quality factor can be obtained as a function of the Boltzman k constant, the absolute temperature T in kelvins, the excess of noise introduced by the negative resistance amplifier A, the VCO output frequency in rad/s ω_0, the specification of the phase noise $L_{1\text{Hz}}$ in dBc/Hz, the value of the capacity of the LC-tank integrated varactor C, the offset frequency with respect to the carrier from which phase noise is evaluated $\Delta\omega$, and the differential amplitude of the VCO output V_A in V.

$$Q = \frac{2 \cdot k \cdot T \cdot (1 + A) \cdot \omega_0}{L_{1\text{Hz}} \cdot V_A^2 \cdot C \cdot \Delta\omega^2} \tag{6.9}$$

From this last equation and taking into account the specification of the phase noise of −110 dBc/Hz and the fact that the minimum amplitude of the VCO output must be 0.5V as stated in Section 6.4.2.4, it is found that the minimum quality factor of the tank necessary in order to achieve this specification of noise must be equal to 6.3. This value has been calculated for a varactor capacity of 0.8 pF and an excess of noise from the amplifier of 2. These first values have been selected due to the fact that for the working frequency of 3.2 GHz, an inductance of 3 nH, and a capacitance of 0.8 pF are common values.

After an inspection of the inductor and varactor library available in the Design Kit, there are inductors and varactors with qualities in the order of 12 and 50, respectively, at 3.2 GHz. Therefore, we can conclude that the initial values selected for the first iteration are consistent with the possibilities offered by the technology used.

6.4.3 Phase Detector

The specifications that are presented in this section are the current value of the charge pump and the noise $L_{1\text{Hz}}$ of the whole component. Despite not depending directly on this block, it is necessary to remember the value of the constant that permits the estimation of the spurs based on leakage. As mentioned in Section 2.2.2.1 this constant called *BaseLeakageSpur* is universal and has a value of 16 dBc.

The preliminary simulations carried out with Simusyn show that the spurs based on leakage are negligible compared to pulse-spurs. This is due to the relatively high-reference frequency of the PLL, 20 MHz in this example. For typical values of the constant *BasePulseSpur*, the current leakage should be in the order of several μA for both types of spurs to have similar values. Evidently, the expected value for the current leakage is much smaller. Therefore, this is not a critical input parameter.

Given that the spurs based on mismatches of the charge pump are the dominant reference emissions it would appear to be logical to include a specification of the constant *BasePulseSpur* in this section. However, given that this constant is difficult to predict, its specification has been determined using Simusyn, assuming the initial requirement of the reference spurs (Section 6.1.3). In addition, in order to obtain this specification it is necessary to know the bandwidth of the loop, of which discussion is presented in Section 6.4.5.

6.4.3.1 Current of the Charge Pump

The influence of diverse factors has been considered in the determination of the current of the charge pump:

- The accuracy of the loop, understood to be the variation of instantaneous voltage at the output from the filter following the introduction a current pulse of a determined amplitude and width in its input;
- The power consumption;
- The value of the capacity placed before the VCO control input (C3 in Figure 6.3). This capacity must be at least three times superior to the input capacity of the VCO line of control.

The accuracy of the loop is not critical for the determination of the current from the charge pump. It is straightforward to verify that the speed of variation in the output voltage of the filter is the same for two input current pulses of equal width but distinct amplitude. This assertion is always correct as long as the loop filter is redesigned each time the value of the amplitude of the input pulse is changed.

From the point of view of the power consumption, it is clear that the ideal situation would be to decrease the current from the charge pump as much as possible. However, in accordance with the expressions in Table 3.3, a reduction of K_ϕ means a reduction in the capacity placed immediately before to the VCO line of control (C3). Therefore, there is a trade-off between power consumption and the VCO control requirements. Finally, the current of the charge pump for the example presented here has been fixed at 50 μA.

6.4.3.2 Phase Detector Noise Normalized at 1 Hz (L_{1Hz})

In Section 5.5 two models of phase noise in the PFD were presented. As with the other building blocks we have adopted the Banerjee model, which uses the empirical parameter L_{1Hz}. The method followed has been the same as in previous cases, with an initial assumption for this variable (–200 dBc/Hz) based on references, and a subsequent correction using the data provided by circuit simulation.

6.4.4 Frequency Divider

Once the value of the output and reference frequencies of the PLL are set it is possible to define the modulus of the frequency divider. In the example presented here the value of the modulus of this divider is fixed at 160.

In reference to its phase noise, it has already been mentioned in Section 4.6 that it is usually negligible compared to the noise that the phase detector introduces inside the loop. This assumption must be verified afterwards, using the electrical simulation tool once all the system can be simulated together.

6.4.5 Global Specifications of the Loop

The last step in the process of defining the starting design values is the determination of global PLL specifications. The values defined here are the extra attenuation of the reference spurs *ATTEN*, the margin of phase ϕ_p, the bandwidth of the loop f_c and the constant *BasePulseSpur*.

Initially, a value of 20 dB has been chosen for *ATTEN*, which may be readjusted once the reference spur level is obtained from the initial simulation of the system. Logically this value of *ATTEN* can be modified providing more or less attenuation to the spurious emissions as required.

Regarding the phase margin, the typical values commonly found vary between 40° and 70°. This parameter is an indicator of the stability of the loop and for this project a value of 55° has been selected. Further information can be found in Section 8.1 of [25].

As in the case of *ATTEN*, this value can be modified by changing the values of the filter components.

With the determination of all the individual specifications of each block and of the global ones, it is then possible to use the Simusyn tool to estimate the bandwidth loop (BW_{LOOP}). As previously explained for this calculation it is necessary to take into account the requirement of the integral of total phase noise from the PLL over the bandwidth of the channel.

In order to clarify the method used in the determination of the bandwidth loop, the inputs used with Simusyn are presented in Table 6.2. These have

Table 6.2
Input Parameters Selected to Specify BW_{LOOP}

Parameter	Value
N	160
R	1
F_{xtal}	20 MHz
$ATTEN$	20 dB
K_{VCO}	140 MHz/V
ϕ_p	55°
L_{1Hz}	−200 dBc/Hz
F_{out}	3.2 GHz
K_ϕ	50 μA
$\lvert f_1 - f_2 \rvert$	200 MHz
tol	1 kHz
$BaseLeakageSpur$	16 dBc

been established in the previous sections, except the constant *BasePulseSpur*, whose specification is determined once BW_{LOOP} is determined.

As well as the data contained in this table, the noise masks of from the reference crystal and from the VCO used in the initial simulation are those specified in Sections 6.1.1 and 6.1.2, respectively.

After some iteration with this data, we conclude that a loop bandwidth of 20 kHz is appropriate in order to fulfill the requirement of the integral of total phase noise from the PLL over the channel bandwidth (20 MHz).

After this point, only the constant *BasePulseSpur* remains to be determined. The estimation of the maximum value that this constant can reach has been undertaken using the Simusyn tool, from the specification of the bandwidth of the loop (20 kHz) and from the requirement of the reference spurious emissions (−59 dBc). Taking into account these values it has been found that the level of *BasePulseSpur* must be less than −278 dBc. This constant must be also recalculated with the data provided by circuit simulation.

Once the bandwidth of the loop and the constant *BasePulseSpur* have been determined, it is necessary to verify that the levels of the phase noise, spurious emissions, and lock time meet the initial requirements presented in Section 6.1. For this purpose, Figure 6.6 shows the total phase noise from the PLL estimated with Simusyn.

With this figure it can be noted that the phase noise simulated meets the requirements previously established; that is, the phase noise from the PLL at the offset frequency of 1 MHz is −110 dBc/Hz and its total integral is less than −32 dBc.

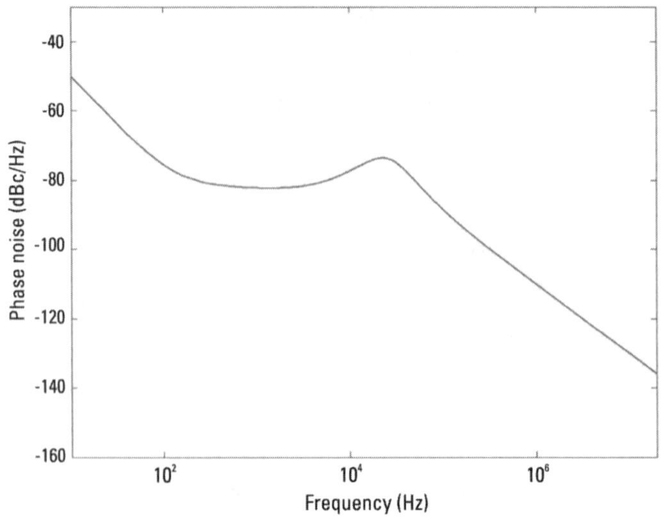

Figure 6.6 PLL phase noise calculated with Simusyn.

As well as the phase noise, the requirement of reference spurious emissions must also be met. For values of *ATTEN* and *BasePulseSpur* of 20 dB and −278 dBc, respectively, with Simusyn an estimated level of reference spurs of approximately −59.2 dBc has been obtained, which is less than the desired requirement (−59 dBc). It is important to point out that the total level of reference spurs has been calculated using the worst case value of the *BasePulseSpur* constant in order to fulfill the requirement of these emissions.

Finally, despite the fact that the frequency synthesizer implemented here in the example presents a fixed output frequency, it is also useful to estimate the lock time of the loop to verify that it is inferior to that required by the standard and that the determination of the specifications of the synthesizer is totally coherent. Using this method, considering the biggest frequency jump possible (200 MHz) and a tolerance of 1 kHz, the lock time simulated is 0.15 ms, which is less than that required by the standard.

So it can be confirmed that the results obtained by means of Simusyn meet the requirements established in Section 6.1, with which it is now possible to begin the design phase of each one of the blocks that constitute the frequency synthesizer.

References

[1] Zargari, M., et al., "A Single-Chip Dual-Band Tri-Mode CMOS Transceiver for IEEE 802.11a/b/g Wireless LAN," *IEEE Journal of Solid-State Circuits*, Vol. 39, No. 12, December 2004, pp. 2239–2249.

[2] Zhang, P., et al., "A Single-Chip Dual-Band Direct-Conversion IEEE 802.11a/b/g WLAN Transceiver in 0.18-μm CMOS," *IEEE Journal of Solid-State Circuits*, Vol. 40, No. 9, September 2005, pp. 1932–1939.

[3] Vassiliou, I., et al., "A Single-Chip Digitally Calibrated 5.15-5.825-GHz 0.18-μm CMOS Transceiver for 802.11a Wireless LAN," *IEEE Journal of Solid-State Circuits*, Vol. 38, No. 12, December 2003, pp. 2221–2229.

[4] Behzad, A. R., et al., "A 5-GHz Direct-Conversion CMOS Transceiver Utilizing Automatic Frequency Control for the IEEE 802.11a Wireless LAN Standard," *IEEE Journal of Solid-State Circuits*, Vol. 38, No. 12, December 2003, pp. 2209–2220.

[5] Krishnapura, N., and P. R. Kinget, "A 5.3-GHz Programmable Divider for HiperLan in 0.25-μm CMOS," *IEEE Journal of Solid State-Circuits*, Vol. 35, No. 7, July 2000, pp. 1019–1024.

[6] Svelto, F., S. Deantoni, and R. Catello, "A 1.3 GHz Low-Phase Noise Fully Tunable CMOS LC VCO," *IEEE Journal of Solid-State Circuits*, Vol. 35, No. 3, March 2000, pp. 356–361.

[7] Park, C., O. Kim, and B. Kim, "A 1.8 GHz Self-Calibrated Phase-Locked Loop with Precise I/Q Matching," *IEEE Journal of Solid-State Circuits*, Vol. 36, No. 5, May 2001, pp. 777–783.

[8] Dec, A., and K. Suyama, "A 1.9-GHz CMOS VCO with Micromachined Electromechanically Tunable Capacitors," *IEEE Journal of Solid-State Circuits*, Vol. 35, No. 8, August 2000, pp. 1231–1239.

[9] Hung, C. M., and K. O. Kenneth, "A Packaged 1.1GHz CMOS VCO with Phase Noise of −126 dBc/Hz at a 600 KHz Offset," *IEEE Journal of Solid-State Circuits*, Vol. 35, No. 1, January 2000, pp. 100–103.

[10] Väänänen, P., M. Metsänvirta and N. T. Tchamov, "A 4.3-GHz VCO with 2-GHz Tuning Range and Low Phase Noise," *IEEE Journal of Solid-State Circuits*, Vol. 36, No. 1, January 2001, pp. 142–145.

[11] De Muer, B., et al., "A 2-GHz Low-Phase-Noise Integrated LC-VCO Set with Flicker-Noise Upconversion Minimization," *IEEE Journal of Solid-State Circuits*, Vol. 35, No. 7, July 2000, pp. 1034–1038.

[12] Grau, G., et al., "A Current-Folded Up-Conversion Mixer and VCO with Center-Tapped Inductor in a SiGe-HBT Technology for 5 GHz Wireless LAN Applications," *IEEE Journal of Solid-State Circuits*, Vol. 35, No. 9, September 2000, pp. 1345–1352.

[13] Zanchi, A., et al., "A 2-V, 2.5-GHz, −104 dBc/Hz at 100 KHz Fully Integrated VCO with Wide-Band Low-Noise Automatic Amplitude Control Loop," *IEEE Journal of Solid-State Circuits*, Vol. 36, No. 4, April 2001, pp. 611–618.

[14] Liu, T. P., and E. Westerwick, "5-GHz CMOS Radio Transceiver Front-End Chipset," *IEEE Journal of Solid-State Circuits*, Vol. 35, No. 12, December 2000, pp. 1927–1933.

[15] Samavati, H., H. R. Rategh, and T. H. Lee, "A 5-GHz CMOS Wireless LAN Receiver Front End," *IEEE Journal of Solid-State Circuits*, Vol. 35, No. 5, May 2000, pp. 765–772.

[16] Behbahani, F., et al., "A 2.4-GHz Low-IF Receiver for Wideband WLAN in 0.6-μm CMOS-Architecture and Front-End," *IEEE Journal of Solid-State Circuits*, Vol. 35, No. 12, December 2000, pp. 1908–1916.

[17] Shana'a, O., I. Linscott, and L. Tyler, "Frequency-Scalable Si-Ge RF Front-End Design," *IEEE Journal of Solid-State Circuits*, Vol. 35, No. 6, June 2001, pp. 888–893.

[18] Wireless LAN Medium Access Control (MAC) and Physical Layer (PHY) specifications: High-speed PHY in the 5 GHz Band; IEEE Standard 802.11a-1999, 2000.

[19] Armada, A. G., "Understanding the Effects of Phase Noise in Orthogonal Frequency Division Multiplexing (OFDM)," *IEEE Transactions on Broadcasting*, Vol. 47, No. 2, June 2001, pp. 153–159.

[20] Bhattacharjee, J., et al., "A 5.8 GHz Fully Integrated Low Power Low Phase Noise CMOS LC VCO for WLAN Applications," *2002 IEEE MTT-S Int. Microwave Symp. Dig.*, Vol. 1, February 2002, pp. 585–588.

[21] Côme, B., et al., "Impact of Front-End Non-Idealities on Bit Error Rate Performances of WLAN-OFDM Transceivers," *Proc. RAWCON*, 2000, pp. 91–94.

[22] Razavi, B., *Design of Analog CMOS Integrated Circuits*, New York: McGraw-Hill, 2001.

[23] Quemada, C., et al., "A CMOS Frequency Synthesizer with Self-Biasing Current Source for 5 GHz Wireless-LAN Receiver," *Microwave Journal*, February 2007.

[24] Agilent Technologies "Advanced Phase Noise and Transient Measurement Techniques," Application Note 5989-1617EN, October 2004.

[25] Caverly, R., *CMOS RFIC Design Principles*, Norwood, MA: Artech House, 2007.

7

Design of a 3.2-GHz CMOS VCO

The objective of this chapter is to present the design of the integrated oscillator that forms part of the frequency synthesizer developed in this book. The specifications established in Chapter 6 and for this building block are presented in Table 7.1.

In the following sections, a suitable architecture from among all those presented in Chapters 2 and 3 is selected, as well as the optimum output stage. Next, the choices of the LC-tank components are discussed. Finally, the design of the oscillator includes its experimental results.

7.1 Choice of Architecture of the Oscillator

In this section the reasons that justify the choice of the most suitable architecture to implement the frequency synthesizer oscillator are outlined.

As analyzed in Chapter 3, the LC-tank integrated oscillator consists of a resonant circuit, an active circuit, and a positive feedback network. In the

Table 7.1
VCO Specifications

Frequency of Output (GHz)	Range of Frequency (MHz)	Amplitude of Output (V)	Phase Noise at 1 MHz (dBc/Hz)
3.2	322	≥0.5	<−110

following sections the reasons that justify the choice of the architectures of each one of these elements are explained.

7.1.1 Tank Circuit

The resonant circuit of the oscillator consists of an inductive (an integrated inductor) and a capacitive component (an integrated varactor). Furthermore, following the discussion shown in Section 3.3.1, the mirror configuration is the most appropriate for the integration of a differential oscillator.

Regarding the varactor, from among the distinct types of varactors presented in Section 3.3.4, the MOS varactors and the triterminals have been rejected due to their excessively abrupt variation in capacitance. The PN junction varactor with finger geometry has been chosen for its larger tuning range compared to those of islands and matrix for very similar quality factors.

Therefore, to implement the LC-tank, two inductors in mirror configuration and two PN union integrated varactors with digit geometry have been selected. The design of these components will be presented in Section 7.2.

7.1.2 Active Circuit

There are two main decisions concerning the active circuit design: the global architecture selection and the transistor type.

7.1.2.1 Architecture Selection

As previously presented in Chapter 3, one of the first classifications of the architecture of the active circuit differentiates between differential or single-ended configuration. In this chapter the differential architecture has been selected for the following reasons:

- It presents numerous advantages compared to the single-ended configuration, despite the fact that as much of the consumption as the occupied area is duplicated. See Section 3.2 for more details.
- The architecture used in the implementation of the mixer *MIX2* (see Figure 6.1) corresponding to the same stage of conversion is also differential.

7.1.2.2 Type of Transistor

Once the differential architecture has been chosen, it is necessary to choose between three types of configurations depending on the use of NMOS, PMOS, or both transistors. It has been described in Section 3.2 that the PMOS architecture is not often used in the design of differential oscillators due to the fact that they present some very clear disadvantages with respect to the other two architectures.

In this way, only two possible configurations are left. The fundamental advantages of NMOS architecture are:

- It permits working with low supply voltages given that it only needs to bias two transistors in saturation, on the amplifier and the current source.
- It introduces less parasitic capacitance than the CMOS topology as the number of transistors included is smaller. This means that for the same values of passive components, the NMOS architecture presents a greater tuning range and a higher frequency of oscillation.

At the same time, the CMOS architecture presents the following advantages:

- It presents a better ratio between phase noise and power consumption. In addition, the fact that there is twice the number of transistors in the active circuit hardly affects the total area of the design, given that it is determined mainly by the size of the resonant circuit.
- It requires less bias current than in the NMOS topology in order to achieve a suitable transconductance value.

Following preliminary circuit simulations, it has been verified that the two architectures permit working at the operating frequency of 3.2 GHz, but that the CMOS configuration does have more difficulties to meet the tuning range of 322 MHz specified in Table 7.1 because of the extra fixed capacitances added by the PMOS transistors. It is for this reason that the NMOS architecture has been chosen as the optimum topology, despite presenting a worse relation between phase noise and power consumption.

7.1.3 Output Stage

The most important requirement of the output stage is that the output from the tank of the oscillator must be perfectly isolated. It is also important to reduce the output impedance so that the gain in voltage is not affected by the subsequent load of 50Ω (required for characterization). In addition, its design must be carried out while attempting to keep the noise added to a minimum and ensuring that the losses from this stage are also kept low. Obtaining this must be attempted using the minimum power consumption possible. These characteristics can be achieved with a common drain stage. In Figure 7.1 a simplified diagram of this circuit can be seen. It should be noted that this

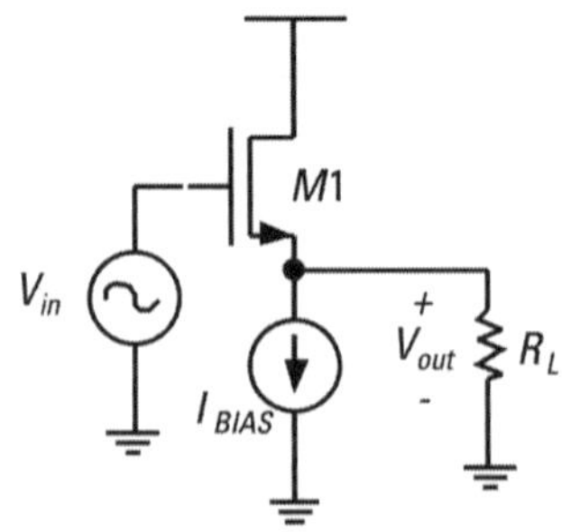

Figure 7.1 Schematic diagram of the common drain stage.

circuit is for measurement purposes only and is not used in the final integrated implementation.

7.2 Design of the Oscillator

After the selection of the architecture and the output stage of the oscillator, this section presents the design process of this component from the schematic circuit to the implementation of the layout of the device.

7.2.1 Basic Expressions for the Design of the VCO

From the circuital diagram in Figure 7.2(a) and substituting its transistors for the high frequency linear model, the equivalent circuit of Figure 7.2(b) is obtained, where R_p represents the resistive losses of the tank, g_{mN} and g_{dN} refer to the transconductance and the admittance from the channel of the NMOS

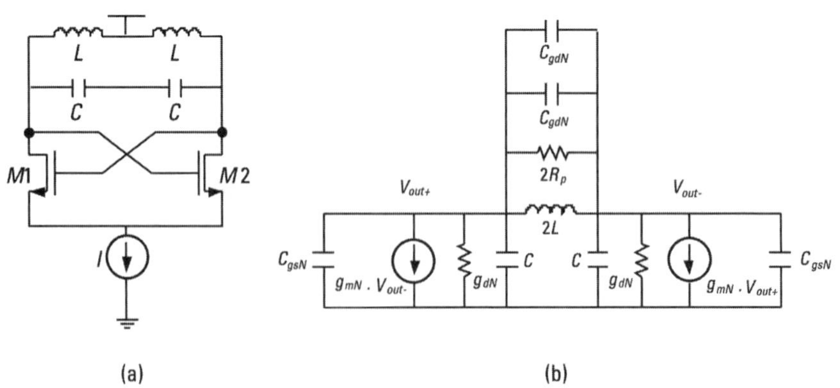

Figure 7.2 (a, b) Circuit diagram equivalent to high frequency from the NMOS oscillator.

transistors, respectively, and C_{gdN} and C_{gsN} consider the capacitance between gate and drain and between gate and source of these transistors. It is important to point out that both NMOS transistors are considered to be identical. In this case, it can be asserted that the output V_{out-} is the same as V_{out+}, but with a phase difference of 180°.

The transconductance of the transistors NMOS (g_{mN}) from the active circuit is given approximately by the expression in (7.1), where K' is a constant fixed by the manufacturing process, W is the width of the channel, L_{eff} is the effective length of the channel, and I_{BIAS} is the bias current.

$$g_{mN} = \sqrt{\frac{2 \cdot K' \cdot I_{BIAS} \cdot W}{L_{eff}}} \tag{7.1}$$

From the circuital diagram equivalent to the one in Figure 7.2(b) and considering the previous assumptions, the value of the admittance Y_{osc} seen from the nodes V_{out+} and V_{out-} can be obtained using (7.2).

$$Y_{osc} = \frac{(C_{gsN} + C + 4 \cdot C_{gdN}) \cdot s + g_{dN} - g_{mN} + (s \cdot L)^{-1} + R_p^{-1}}{2} \tag{7.2}$$

From this admittance the condition and the frequency of oscillation can be calculated using (7.3) and (7.4), respectively.

$$R_p^{-1} + g_{dN} < g_{mN} \tag{7.3}$$

$$f_{osc} = \frac{1}{2 \cdot \pi \cdot \sqrt{L \cdot [C + 2 \cdot (C_{gdN} + C_{gdN}) + C_{gsN}]}} \tag{7.4}$$

where L is the inductance of the inductor and C is the capacitance of the varactor.

In accordance with (7.4), the oscillator will function in steady state as long as the value of the real part of the admittance of the active circuit is greater than the resistive losses from the tank.

In (7.4) it can be seen that the oscillation frequency depends on the value of the parasitic capacitances between gate and drain and between gate and source, from the transistors of the active circuit, which at the same time depends on the size of these transistors. As well as influencing the frequency of oscillation, these parasitic capacities can reduce the range of frequency covered by the VCO.

By inspection of (7.3), it can be deduced that in order to ensure oscillation it is recommendable to increase the transconductance of the NMOS transistors as far as possible. This must be carried out taking into account the influence of the resistance of the channel of the transistors in the value of the total admittance. Furthermore, it is important to point out that an increase in the transconductance g_{mN} of both transistors is achieved by increasing the current of polarization or the width of the channel of these. An increase in the dimensions of the transistors of the active circuit involves an increase in its parasitic capacities and as a consequence a reduction in the maximum frequency and in the frequency range of the VCO.

7.2.2 Design and Selection of the Tank Circuit

The tank circuit plays a key role in the behavior of the integrated oscillator. In addition, as seen in the previous section, the design of the active stage depends on the resistive losses from this tank.

The difficulty of this point arises from the traditionally poor models of passive components, specially the inductors. In many currently available design kits, the foundries offer dedicated high frequency modeled components, which certainly was not the case some years ago. Even if RF models are available, when we face a new design, we recommend carefully reviewing these models before entering further into the design process. But, without any doubt, the safest strategy is to develop our own design kit despite of the extra cost to do it.

This is the path we followed in this book, where the inductors and varactors were characterized and modeled as shown in the following sections.

7.2.2.1 Design and Selection of the Integrated Inductor

As a result of a quick circuit analysis, an inductor of 2 nH has been chosen for the tank. This leads to a reasonable varactor, consistent with tuning range, power consumption, and occupied area requirements. Therefore, the task now is to develop a microlibrary of inductors of this value, optimized to work at 3.2 GHz.

For this purpose, a set of four inductors has been designed using an electromagnetic simulation tool: ASITIC (Analysis and Simulation of Spiral Inductors and Transformers for ICs), developed at Berkley [1]. Subsequently, these inductors have been implemented with the CMOS 0.18 μm process and characterized. In Table 7.2 the experimental results are presented. The parameters shown are the maximum quality, the frequency that this quality has reached, and the values of inductance at frequencies of 2.4 and 5 GHz.

These inductors have been implemented with the top metal layer, whose thickness is 2 μm (named as thick metal 6), and the geometry has been optimized in order to obtain the best quality factor. The method followed is not explained here, since it is not part of the objectives of this book, but we encourage the reader to find more information in [2].

Table 7.2
Experimental Results of Inductors in CMOS 0.18 Technology

Inductor	*L* at 2.4 GHz (nH)	*L* at 5 GHz (nH)	Q_{max}	*f* at Q_{max} (GHz)
L1	1.3	1.4	10.8	6.5
L2	1.8	1.9	10.1	4.9
L3	2.1	2.3	9.9	3.9
L4	2.5	2.8	9.7	3.9

The inductor most appropriate for this work is L3 from Table 7.2 because it combines Q_{max} of 10 at a frequency close to the center frequency of the VCO. This component has 2.5 turns, 20 μm of track width, and 160 μm of diameter. The diameter is twice the smallest interior radius of the inductor. Figures 7.3 and 7.4 show the experimental results of the inductance and the quality factor versus frequency, showing the peak Q value at the desired VCO operating frequency.

Figure 7.5 demonstrates a microphotograph of the integrated inductor designed including the structure for its characterization.

In order to simulate the inductor with the rest of the circuit, the results obtained from the laboratory have been used to obtain an Π-model shown in Figure 7.6. The approximation error is less than 0.5%. The characterization of

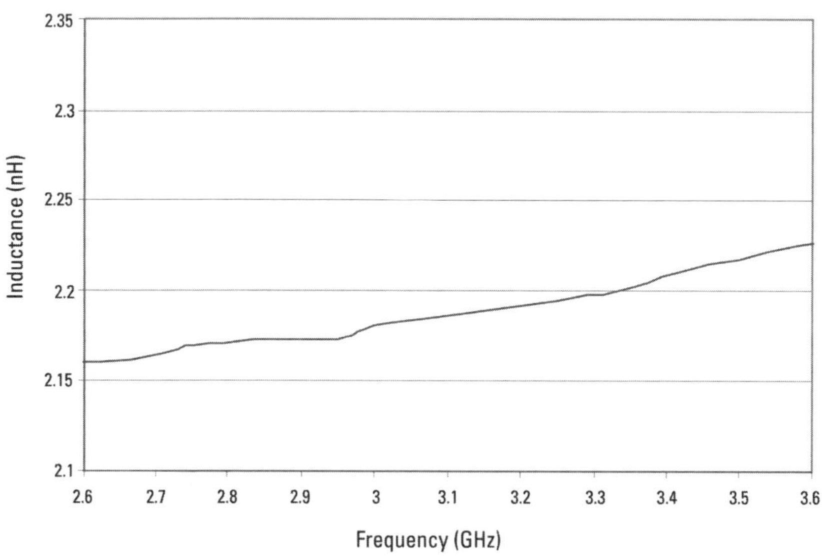

Figure 7.3 Measured inductance of the integrated inductor versus frequency.

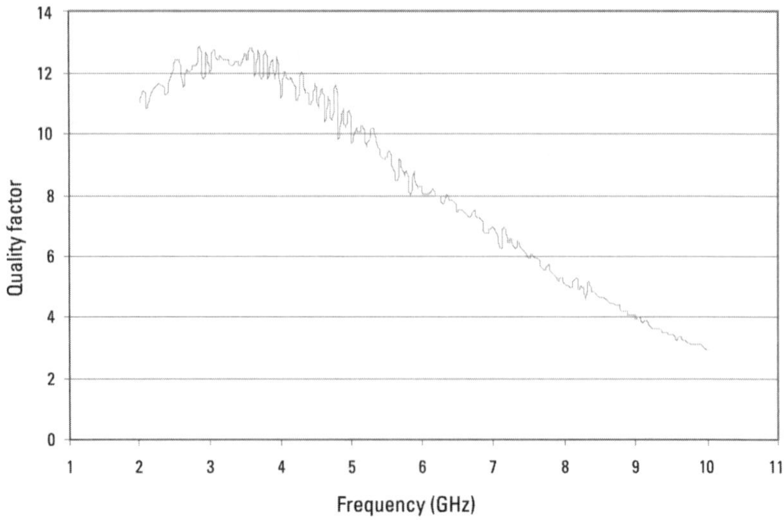

Figure 7.4 Measured quality factor of the inductor versus frequency.

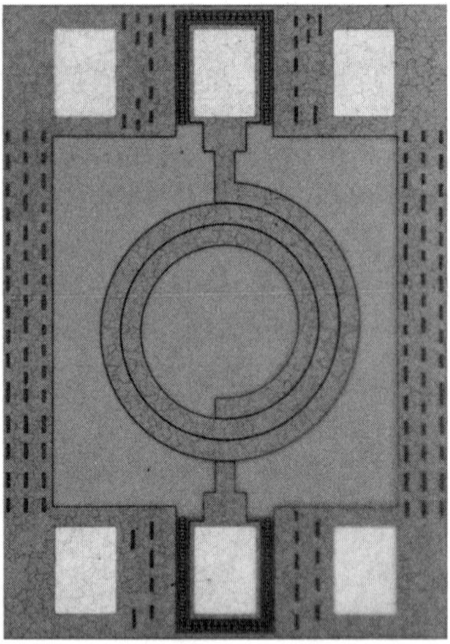

Figure 7.5 Microphotograph of the inductor designed.

this inductor has been performed with IC-CAP (Integrated Circuits Characterization and Analysis Program) from Agilent, and the error comparison with the iterative automatic tool "optimization" in ADS also from Agilent.

Therefore, this model will be the one used in the design of the NMOS integrated oscillator used in the example for this book.

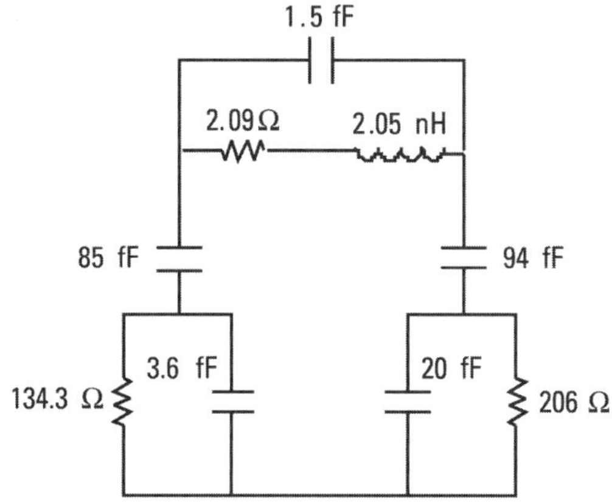

Figure 7.6 Adjusted model in Π of the inductor designed.

7.2.2.2 Design and Selection of the Integrated Varactor

Once the inductor has been designed, it is necessary to select the integrated varactor. As already seen in (7.4), the frequency of oscillation of the output of the VCO depends, in addition to the capacitance of the varactor, on the parasitic capacities of the tank. The most important parasitic capacitances are those corresponding to the transistors of the active circuit, although there are also contributions of the substrate of the inductor, the output stage and the tracks of the circuit. In addition to these issues, it is important to take into account the parasitic inductance from the metal track that joins the two mirror inductors. Logically, this parasitic inductance can not be estimated until the layout of the circuit is completed.

Through an initial simulation of the oscillator, it has been estimated that the maximum capacitance of the varactor (capacitance for zero voltage between its terminals) should be nearly 0.9 pF (65 fingers). Later on, this capacitance will have to be reduced when all the parasitic capacitances that affect the tank are taken into account.

Another key input data for the design is the uncertainty in the data provided by the foundry, which has to be incorporated as an "extra" tuning range. This data must be obtained directly from the information supplied in the design kit. In our case, the RMS error of S-parameters of a 60-finger PN varactor can be seen in Table 7.3.

In this table it can be seen that the RMS is always less than 3%. This provides a considerable security in order to achieve the specification of the frequency of output from the integrated oscillator. With the inductor of the tank correctly modeled and with the accurate optimization of the varactor, it is possible to implement the oscillator with high guarantees of obtaining the required frequency of operation.

Table 7.3
RMS Error of the S-Parameters of a 60-Finger PN Union Varactor

RMS_{S11} (%)	RMS_{S12} (%)	RMS_{S21} (%)	RMS_{S22} (%)
1.792	1.371	2.685	2.655

With regard to the quality factor, in Figure 7.7 it is easy to verify that the quality of the varactor is high compared to the inductor. This quality factor has been obtained for a reverse bias voltage of 1.65V and presents a value of 58.7 for a frequency of operation of 3.2 GHz.

The circuital diagram adopted to model the component is illustrated in Figure 7.8, where each component has the following physical meaning:

- L_{C1} and L_{C2} account for the parasitic inductances of the metallic contacts.
- D_{PN} represents the capacitance of depletion of the PN junction.
- C_N refers to the capacitance of the N-well.
- D_N considers the variable capacitance between the well N and the substrate P.

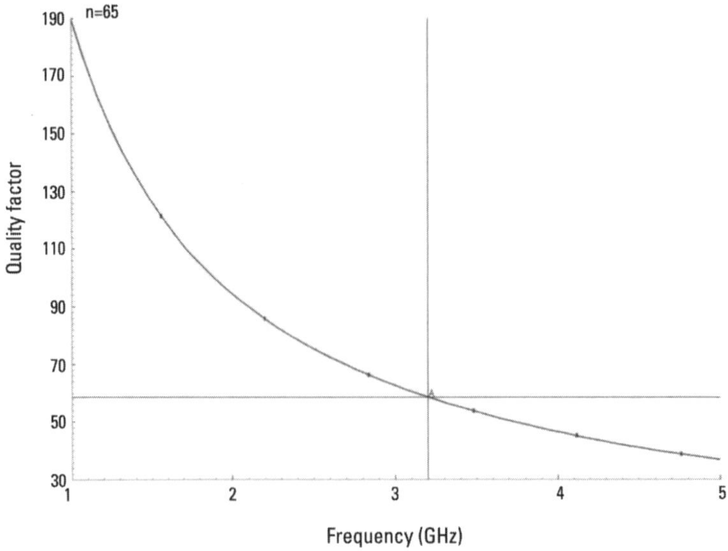

Figure 7.7 Simulated quality factor of the varactor initially selected.

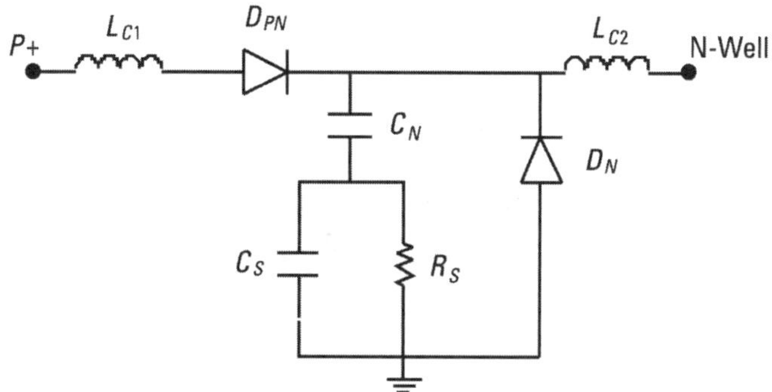

Figure 7.8 Schematic model of the PN junction varactor.

- R_S and C_S represent the resistance and capacitance of the substrate, respectively.

Now the circuital model used has been presented it is useful to point out that the final PN union varactor laid out must present a slightly lower capacitance, due to the influence of all the parasitic capacitances that appear in the tank. As mentioned earlier, it is not possible to determine its values exactly beforehand. Therefore, the previous experience of the designer is vital at this point to estimate the appropriate reduction.

The varactor finally selected has 56 fingers, a capacitance at zero voltage of 758.46 fF, and a quality factor of 58.81 at 3.2 GHz. Figure 7.9 presents a microphotograph of the varactor finally selected, which occupies an area of $10 \times 10\ \mu m^2$, without the measurement structures.

The combination of this varactor and the inductor presented in the previous section gives a total quality factor for the tank higher than 10. This value is consistent with the requirements of the circuit, which has been developed in Chapter 6.

Finally, we can perform an approximate calculation of the frequency range. Logically, the real range of tuning will be less than that estimated as a consequence of the parasitic effects. Nevertheless, this value can be used as an initial guiding reference. Considering the varactor of 56 digits and assuming that the frequency is 3.2 GHz, a tuning voltage of 1.65V is reached, the range of frequency estimated is approximately 808 MHz (for a tuning voltage between 0.5V and 2.8V), way higher than the initial requirement contained in Table 7.1. It is important to highlight that the capacitance of the varactor to the tuning voltages of 1.65V and 2.8V is 506.1 and 442.4 fF.

7.2.2.3 Calculation of the Conductance of the Tank

Following the selection of the two passive components of the tank, it is necessary to calculate its equivalent total conductance in order to be able to find out the

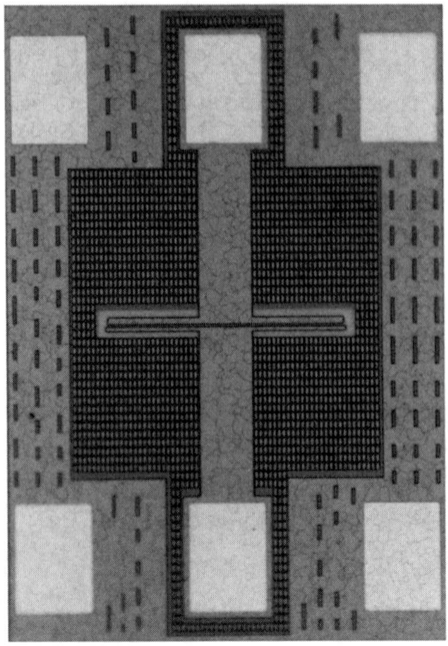

Figure 7.9 Microphotograph of the varactor finally selected.

negative conductance that must be supplied to the active circuit in order to maintain the oscillation in steady state.

Using the simplest model for each one of the two passive elements of the tank (a capacitance or inductance with a series resistor), it is possible to calculate a preliminary estimation of the conductance of the tank. Figure 7.10 illustrates the transformation that must be carried out in order to obtain this estimation, something which can be done using (7.5).

$$G_{eq} = \frac{R_C \cdot (\omega \cdot C)^2}{1 + (R_C \cdot \omega \cdot C)^2} + \frac{R_L}{R_L^2 + (\omega \cdot L)^2} \tag{7.5}$$

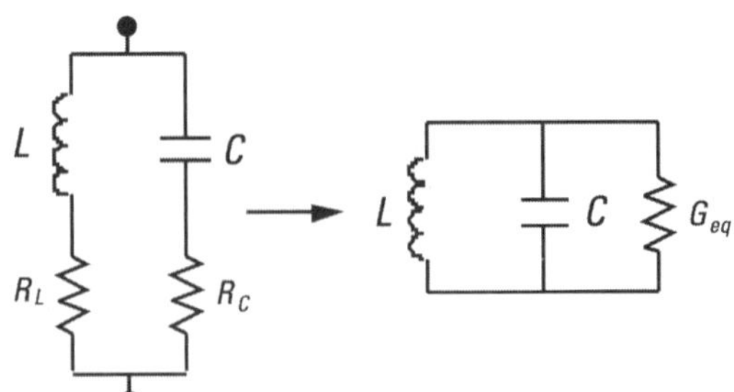

Figure 7.10 Equivalent circuits of the tank.

Through (7.5) and knowing the values of R_C and R_L are 1.439 and 2.02Ω, the conductance equivalent of the tank obtained is 1.435 mS.

In order to obtain a more accurate value, it is necessary to use the circuit simulation tool. By using this method, carrying out an S-parameter analysis to the circuit of Figure 7.11, the conductance of the tank can be calculated as the real part of the admittance seen from the input port.

Figure 7.12 shows the result of the simulation of this conductance, which presents a value of 1.401 mS at the working frequency of 3.2 GHz.

7.2.3 Design of the Schematic Circuit of the Oscillator

The steps followed to complete this design have been:

1. Characterization of the transistors of the active circuit;
2. Current source design;
3. Complete circuit simulation.

The following sections describe each one of these points in detail.

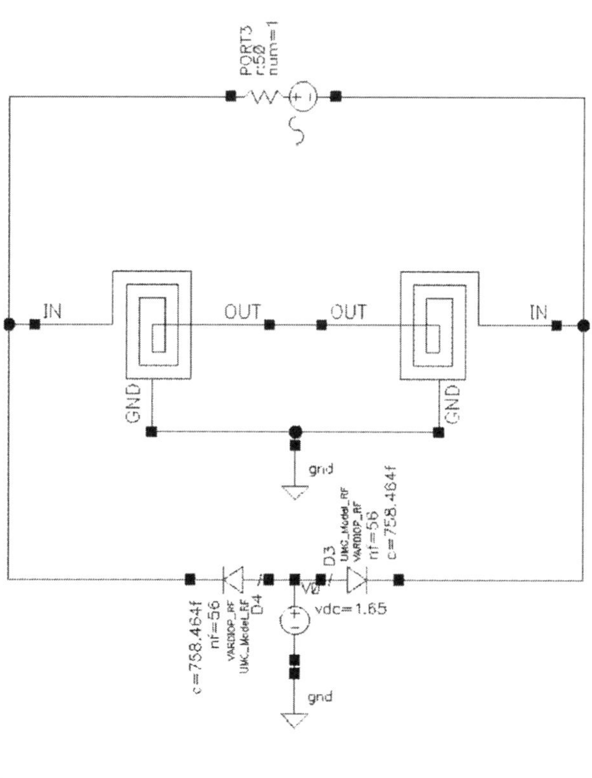

Figure 7.11 Schematic circuit of the tank.

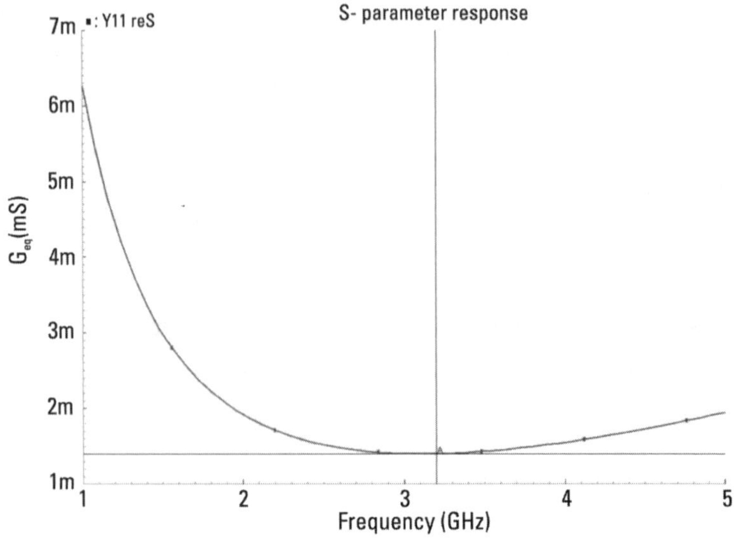

Figure 7.12 Simulation of the conductance of the tank.

7.2.3.1 Characterization of the Transistors of the Active Circuit

Once the passive components of the tank have been selected and the value of its conductance estimated, the active circuit must be designed so that it is capable of compensating the resistive losses in order to be able to maintain a stable oscillation.

From (7.2), and leaving to one side the parasitic capacities from the tank and the admittance from the channel of the NMOS transistors, the admittance that the active circuit generates can be estimated using (7.6).

$$Y_T = -\frac{g_{mN}}{2} \tag{7.6}$$

In accordance with the value simulated in Figure 7.12 and from (7.6), each transistor should generate a transconductance of approximately 2.803 mS in order to ensure the stable functioning of the oscillator.

However, this does not ensure that the oscillator will start to function. In order to secure this aspect, the conductance from the active circuit must be superior to that generated by the tank. Normally a security factor of between 2 and 3 is used.

In order to achieve the greatest transconductance possible, while maintaining the transistor size at a minimum, the length 0.18 μm has been selected. The bias current has been set at 6 mA by means of simulation. With this value

of current a total value of the conductance of 3.67 mS is obtained, which leads to a security factor of 3.62. Evidently, this factor of security will be reduced when substituting the ideal elements for its models and introducing the output stage of the oscillator.

7.2.3.2 Current Source Design

For the first simulations of the resonant circuit together with the transconductance amplifier, the circuit has been biased with an ideal current source. For the definitive implementation, a source compensated by voltage has been selected [3]. This configuration helps to reduce the influence of the variations of the voltage supply. Its diagram together with the point of operation desired can be seen in Figure 7.13, where V_{TN} is the threshold voltage of the NMOS transistor $M1$, V_{GS1} is the voltage between the gate and source of the transistor $M1$, β_1 is the gain of the transistor $M1$ depending on the manufacturing process and on the relation W/L, and I_{D1} and I_{D2} are the drain currents from the transistors $M1$ and $M2$, respectively. The cascode formed by $M3$ and $M4$ help to improve the performance of the circuit given that it increases the output resistance, as well as the Wilson structure current mirror made up by $M1$, $M2$, and R.

In Figure 7.13(a) the desired operation point can be seen (A) together with another possible state of absence of current (B). This constitutes the main disadvantage of these types of sources. In order to solve this problem, it is necessary to use a *startup* circuit that allows the source to abandon the point of operation B until point A has been reached. This circuit (*startup circuit*) is illustrated in Figure 7.13(b) and further explanation is found in [3].

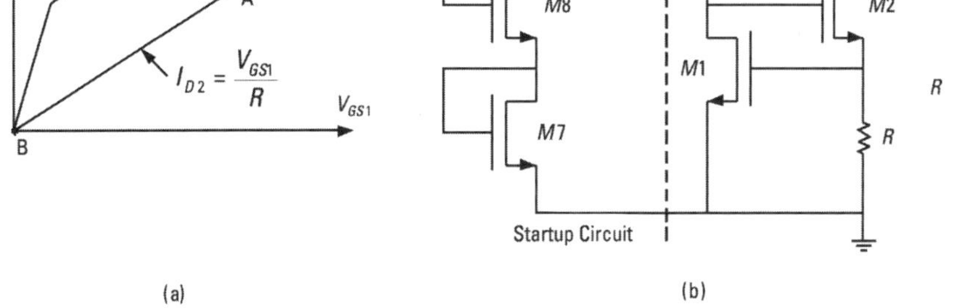

Figure 7.13 (a, b) Diagram of the current source of the active circuit.

Figure 7.14 presents a simulation corresponding to the variation of the output current of the source when the feed voltage of this is varied by 10% of its operating value. In this figure it can be seen that for such a variation the variation of the current is of the order of 3%.

7.2.3.3 Complete Circuit Simulation

Before we perform the final circuit simulation, it is important to introduce a series of elements in the circuital diagram that permit obtaining more accurate results.

First, the capacitor in parallel with the current source must be included, in order to improve the phase noise. After carrying out different simulations of phase noise from the oscillator, an optimum value of 10 pF has been selected and added, with an RF optimized Metal-Insulator-Metal capacitor.

It is also necessary to include the output stage of the oscillator, a common drain stage and to carry out the optimization, basically adjusting transistor size and current to obtain a good combination of power consumption and gain. Note that the parasitic capacitance added to the tank will depend on the transistor size, but also on the current of the buffer, because some of these effects are bias-dependent.

Another factor to take into account is the connection pads. These are metal structures that mainly introduce resistances and capacitances, and its

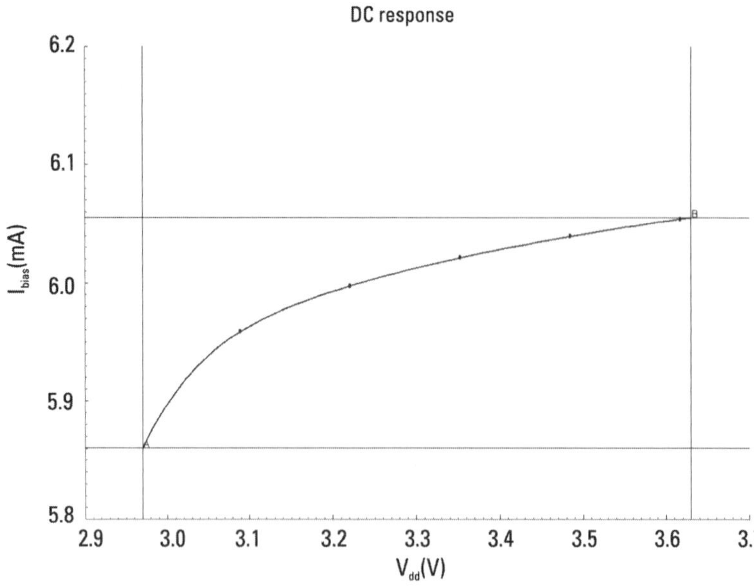

Figure 7.14 Variation of the output current of the bias source of the tank.

structure must be analyzed and modeled. In or case the supply pads are formed by six layers of metal connected to each other by means of matrixes of vias. The ground pads also consist of the six layers of metal, but the bottom metal (Metal 1) it also connected to the substrate. Finally, the RF input/output pads consist of the two most external layers (Metal 6 and Metal 5) connected together, leaving the lower metal without connection.

In addition to the three previous elements, it is useful to introduce capacitors of a high value between the supply voltage and ground in order to filter the noise coupled via voltage sources. The resulting schematic circuit is presented in Figure 7.15.

As a result of the simulations of the complete circuit, the optimum values selected for the principal elements of the circuit in Figure 7.15 are included in Table 7.4.

The results obtained from the simulation of the circuit in Figure 7.15 are presented in Table 7.5.

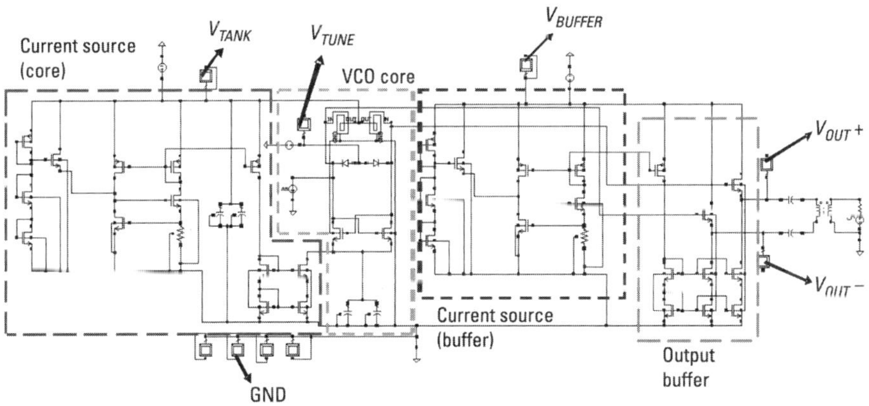

Figure 7.15 Complete circuital diagram of the NMOS oscillator.

Table 7.4
Values of the Principal Variables of the VCO

Channel width of the active circuit transistors (μm)	25
Channel width of the output stage transistors (μm)	25
Channel length of the active circuit transistors (μm)	0.18
Channel length of the output stage transistors (μm)	0.18
Capacitor in parallel with the source (pF)	10
Bias current of the tank circuit (mA)	6
Bias current of the output stage (mA)	6

Table 7.5
Results of the Simulations of the Circuital Diagram of the VCO

Frequency of oscillation for V_{tune} = 1.65V (MHz)	3,383.87
Frequency of oscillation for V_{tune} = 0.5V (MHz)	3,526.16
Frequency of oscillation for V_{tune} = 2.8V (MHz)	3,110.41
Phase noise at 1 MHz (dBc/Hz)	–119.1
Power of output (dBm)	3.8
Power consumption including the output stage (mW)	83
Differential amplitude at the output of the tank (V)	2.32

7.2.4 Layout Implementation

Figure 7.16 presents the layout of the oscillator, whose schematic circuit is shown in Figure 7.15. The main guidelines followed during its design have been pointed out in Section 3.5. As for the passive inductive element, it is advisable that the interconnections of the elements from the tank are carried out by means of the top layer of metal given that they usually have a lesser resistance and minimize the parasitic capacitance to the substrate due to the fact that it is the furthest layer of metal from this.

Figure 7.16 clearly shows nine pads of interconnection with the exterior and two large concentric rings surrounding the design. Pads functions are as follows:

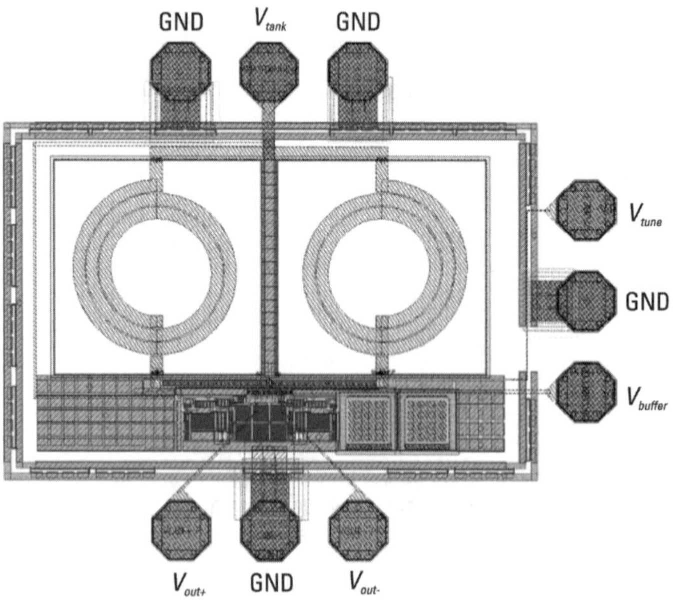

Figure 7.16 NMOS oscillator layout (schematic circuit shown in Figure 7.15).

- V_{tank}: V_{DD} for the core VCO section (LC-tank and G compensation circuitry);
- V_{buffer}: Bias voltage for the output buffer;
- V_{tune}: VCO control line;
- V_{out+} and V_{out-}: VCO differential outputs;
- GND: Ground connections.

Two square rings have been implemented to reject external influences on supply rails. There are also additional rings surrounding the varactors, the inductors, and the transistors, improving the isolation and minimizing the noise coupled through the substrate. Ground contacts have been provided in the substrate throughout the circuit in order to reduce the effect of undesirable couplings. Another feature used in the layout of this example is the integrated capacitors in the order of tens of picofarads between V_{DD} and ground so that the possible noise coupled through the voltage sources finds a low impedance path to ground.

Another feature used in the layout of this example is the integrated capacitors in the order of tens of picofarads between V_{DD} and ground so that the possible noise coupled through the voltage sources finds a low impedance way to ground. These elements can be observed on the right of the active part of the block in Figure 7.16.

Finally, the metallic tracks of interconnection between components have been designed using the two criteria already described in Section 3.5. In the case of the example presented here, a current density of 0.4 mA/μm^2 is a valid estimation for any metal track, but for the higher one which I_{max} is close to 1 mA/μm^2 as it is thicker.

In the final complete design of the PLL (see Chapter 10), the considerations corresponding to the core of the VCO are described. This part of the layout, consisting of the resonant tank and the transistors from the active circuit and from the output stage, is without doubt the part that requires the highest meticulousness in its design. In Figure 7.17 a zoom of the layout of the core of the oscillator is shown.

In the upper part of Figure 7.17 two varactors can be observed, while the inductors would be located above these at both sides. The transistors from the active circuit and from the output stage are situated below the varactors. As can be observed in the layout of the nucleus of the VCO, the maximum degree of symmetry possible has been maintained, given that this is especially important in differential architectures in order to avoid imbalances between branches and, as has already been seen, above all in order to reduce the phase noise. Furthermore, the effect of the possible mismatches from the manufacturing process is also reduced. Although common centroid configurations are often used, in this

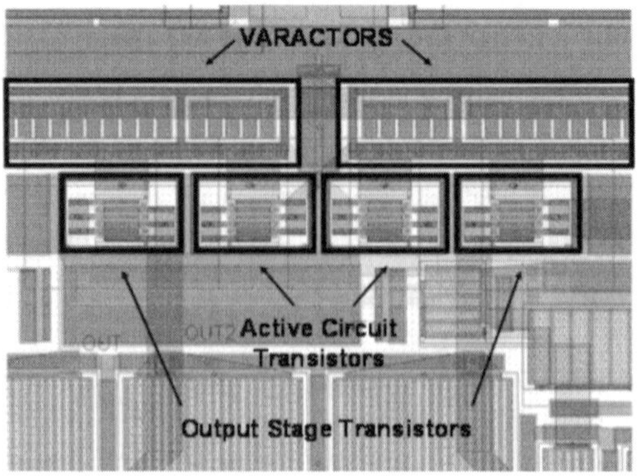

Figure 7.17 VCO core layout (schematic circuit shown in Figure 7.15).

particular example this technique has not been used, due to the minimum size of the transistors.

As previously explained and yet introduced in Chapter 3, the unavoidable amount of parasitic effects makes the redesign of the varactors essential once the layout of the oscillator has been completed and simulated in its post layout version. Variations no higher than 5% should been registered but whose effect in the output parameters can turn out to be dramatic. For example, the connection track between the inductor and the varactors in this case introduce a parasitic inductance on the order of 150 pH per inductor and a parasitic capacitance on the order of 100 fF.

When comparing the results of Table 7.5 with the specifications of the VCO contained in Table 7.1 it can be seen that all the requirements are met except the frequency of oscillation, which presents a slightly superior value to that desired. The final value must be adjusted during the layout definition.

References

[1] ASITIC homepage: http://rfic.eecs.berkeley.edu/~niknejad/asitic.html.

[2] Aguilera, J., and R. Berenguer, *Design and Test of Integrated Inductors for RF Applications*, New York: Springer, 2003.

[3] Gray, P. R., and R. G. Meyer, *Analysis and Design of Analog Integrated Circuits*, New York: Wiley, 1993.

8

Design of a Frequency Divider

The objective of this chapter is to explain the design of the second block of the locked loop type synthesizer used as an illustrative example for the purpose of this book: the frequency divider. As with the previous block, the architecture of the divider selected will be justified, followed by the presentation of the design of this component, and finally the results of the postlayout simulations will be explained. These will be compared with its experimental results in Chapter 11.

As seen in Chapter 6, the most important specification of the frequency divider is the module of division, considering negligible its phase noise with respect to the dominant sources of noise within the bandwidth of the loop; these are the reference crystal and the phase detector. Logically, this will be verified later on in the postlayout simulation of the component.

8.1 Choice of the Architecture of the Divider

The frequency divider used in this example consists of two high-frequency dividers-by-2 and one low-frequency divider of 40 for a total division ratio of 160.

As a consequence of the high frequency of the input signal to the divider, it is necessary to implement two high-frequency dividers-by-2 throughout that convert the signal from 3.2 GHz to 800 MHz. Once this conversion has been carried out, it is possible to use low-frequency digital dividers. This means that special RF design techniques just need to be used in the two high-frequency dividers.

Figure 8.1 shows the diagram of blocks of the complete divider.

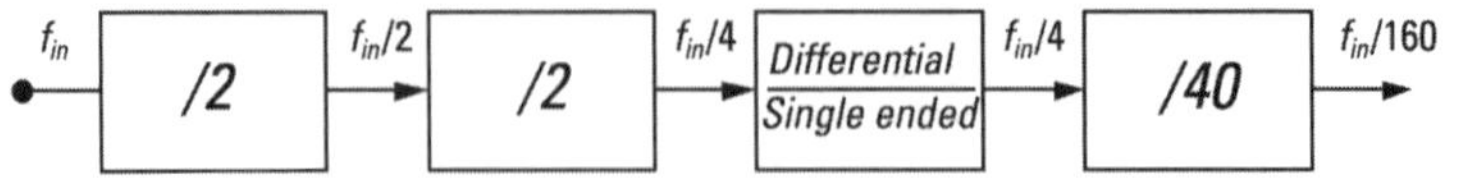

Figure 8.1 Block diagram of the frequency divider.

As described in Section 6.1, the VCO selected for this example is NMOS differential. Due to this, a frequency divider that is capable of operating with these two differential signals must be designed. This is achieved through the two first stages of high-frequency division. However, the logic gates as much as the flip-flops, and in general the low-frequency dividers designed, possess a single-ended architecture. In addition, as these low-frequency digital dividers require square wave signals of amplitudes between 0V and the supply voltage, it is also necessary to digitize the signal at its input for their correct functioning. This is achieved by means of a differential/single-ended converter. Once the frequency has been reduced and the signal converted to single-ended, it is then possible to use a low-frequency digital divider of 40.

In the following sections the architectures of each one of the blocks in Figure 8.1 is presented.

8.1.1 High-Frequency Divider-by-2

The choice of architecture at this stage is, without any doubt, critical for the complete divider designed in this example. This aspect is due mainly to its high operating frequency and to its high power consumption in comparison to the rest of the blocks.

In Chapter 4 the different architectures found for implementing high-frequency dividers-by-2 have been presented. Among them, the SCL with static charges has been chosen for the following reasons:

- It is a faster topology than the other three due to the total absence of PMOS transistors and to the fact that the signal only circulates through two gates per cycle. In addition, for frequencies around 5 GHz and a correct layout of its transistors, the increase of power consumption is not critical.
- The SCL architecture functions correctly for smaller dynamic ranges of input clock than in the other three topologies. This characteristic guarantees a wider safe margin.

Therefore, each high-frequency divider will be formed by two latches in the master/slave configuration (as shown in Figure 4.5), each one of which will be implemented with SCL topology (as shown in Figure 4.9).

8.1.2 Differential to Single-Ended Converter

As described previously, due to the characteristics of the low-frequency digital divider that precedes the division by 2 stages, it is necessary to convert the signal from differential to single-ended.

The circuital diagram used to design this converter is illustrated in Figure 8.2. The main reasons that justify the choice of this architecture are its simplicity and its moderate consumption of power (less than 1 mW according to [1]). The figure of merit of the frequency dividers is the ratio of the power consumption over the working frequency, that is, mW/MHz, and this is why the power consumption is a key parameter in the architecture selection for this kind of blocks, where V_{in} and $\overline{V_{in}}$ are the differential output signals of the second high-frequency stage of division by 2 and V_{out} the output signal of the converter.

This converter is formed by two NMOS differential amplifiers in phase opposition (M5, M6, M9, and M10) that feed two PMOS amplifiers in common source (M3 and M4) connected by means of an NMOS current mirror (M1 and M2). The two NMOS amplifiers must be implemented using symmetric loads (M7, M8, M11, and M12) and the same bias current. These provide signal amplification and DC bias for the PMOS amplifiers in common source, as well as providing an additional amplification to carry out the conversion to a single-ended signal through the current mirror.

8.1.3 Low-Frequency Digital Divider

The operating frequency of this block is the lowest of all the stages that form the frequency divider. Section 4.5 describes how this type of low-frequency digital dividers consists of flip-flops and logic gates, but with a series of fundamental

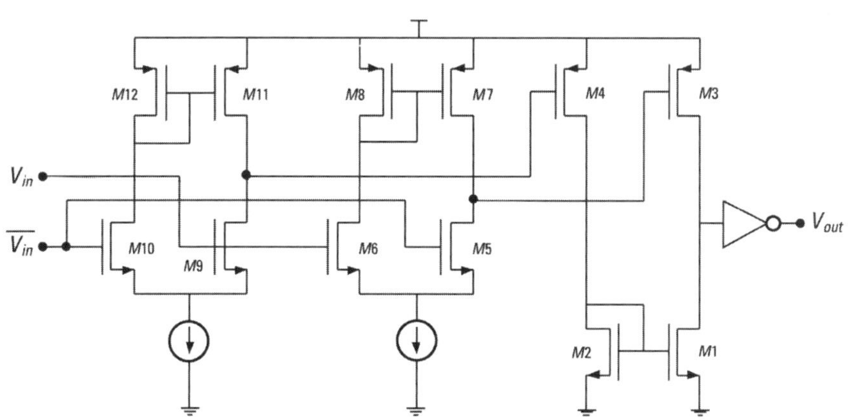

Figure 8.2 Schematic circuit of the differential to single-ended converter.

differences with respect to high-frequency dividers: the configurations used for implementing these flip-flops and logic gates are different and are closer to digital electronic design.

Furthermore, as seen in Section 4.6, the total phase noise from the divider is usually dominated by the phase noise from the high-frequency stages.

In this way, as the value of the module of the low-frequency divider is 40, this block has been divided in a stage of division by 8 and another by 5. Given that the module 8 can be obtained through 2^3, designing a divider by 2 and another by 5 is sufficient.

In Figure 8.3(a, b) the configurations used to develop the digital dividers by 2 and 5 are demonstrated, respectively.

As can be seen in Figure 8.3, these digital dividers are implemented combining type D flip-flops and a logic gate NAND. The configuration of logic gates used to develop each one of the flip-flops of the low frequency digital divider is illustrated in Figure 8.4, where IN1 and IN2 are unclocked and clocked inverters, respectively. The first is conventional, while the second is controlled by two 180° dephased signals, *CLK* and its complement $\overline{CLK}$, in a way that the signal is only inverted when *CLK* is at a high level.

The circuital diagrams selected to develop the three types of logic gates used in the implementation of the low frequency digital divider are presented

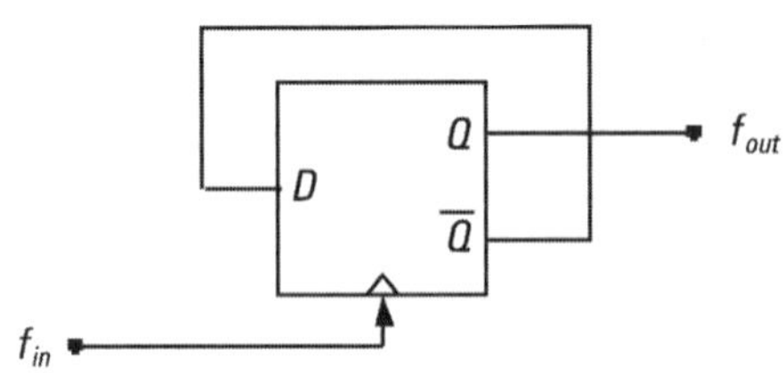

(a)

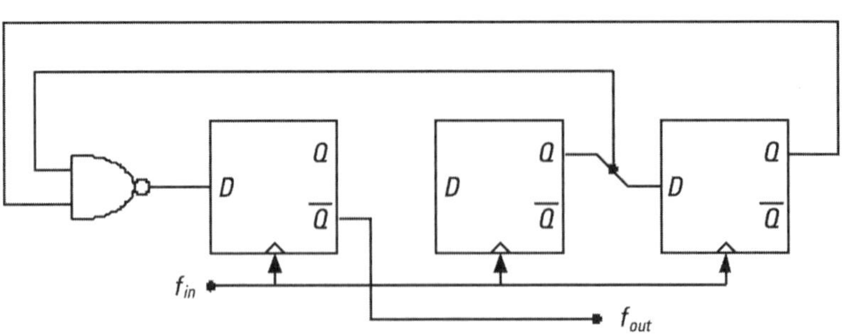

(b)

Figure 8.3 Configurations of the digital dividers: (a) divider-by-2, and (b) divider-by-5.

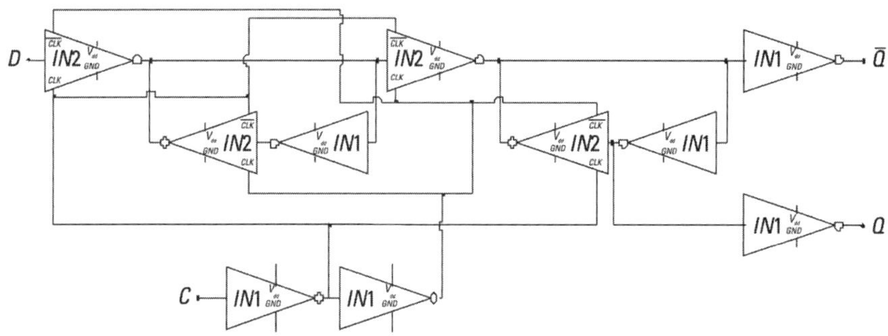

Figure 8.4 Configuration of logic gates of D flip-flops

in Figure 8.5. The transistors that are part of these digital logic gates normally are, unless special driving requirements, of their minimum size.

The next section shows the design of the complete divider using these circuits as building blocks.

8.2 Design of the Frequency Divider

This section presents the process of designing each one of the blocks of the frequency divider until the implementation of the complete layout of this device is obtained.

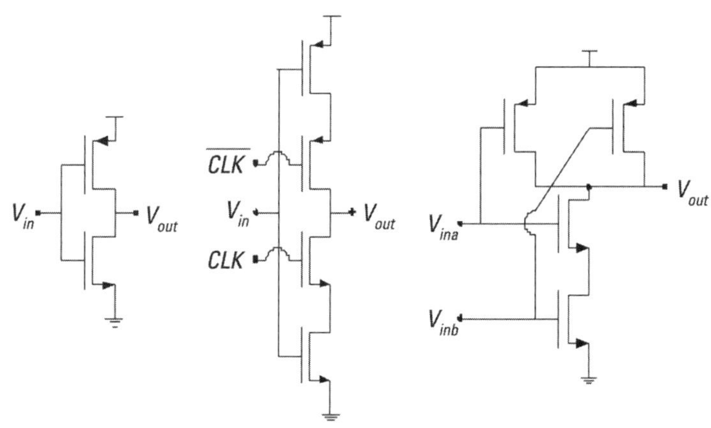

(a) (b) (c)

Figure 8.5 Circuital diagrams of the logic gates: (a) IN1, (b) IN2, and (c) NAND.

First, the design of the two divider-by-2 stages and the converter are explained. Next the characterization of the circuit is presented. Finally, the layout of this component will be presented.

8.2.1 High-Frequency Divider-by-2 and Converter

The objective of this section is to present the design of the two stages of high-frequency dividers together with the single-ended to differential converter. In Chapter 4, and concretely in (4.2), the dependence of the jitter of the divider by 2 with SCL topology with the output capacitance was demonstrated. The input capacitance of the following stage, as well as the divider-by-2, contributes to the value of this capacitor. Due to this, the presentation of the design together with these blocks has been considered suitable.

In accordance with the study developed in Section 4.6 regarding the phase noise that a high-frequency flip flop with SCL topology introduces, the main source of noise is due to the PMOS transistors that act as resistors biased in the triode region. In the particular case of this example, these PMOS transistors have been directly substituted by resistors. This analysis makes use of (4.2) to estimate the noise introduced by these resistors and generated by each flip-flop of the divider. Through this expression it can be said that this noise is directly proportional to C_T and inversely proportional to the squared bias current. It is important to point out that C_T is the total output capacitance of the flip-flop without load plus the capacitance of interconnection and the input capacitance of the following stage.

Therefore, to minimize C_T in each flip-flop and, in short, the phase noise introduced by the first two high-frequency stages, transistors of minimum size have been selected.

Once the active components have been chosen, the resistors and the bias sources of each latch can be defined. Basically, this choice must observe a compromise between phase noise and power consumption, taking also into account the frequency at which each flip-flop must function and the required input amplitude. Following an iterative process with the circuit in Figure 8.6, based on optimizing each one of its blocks of high frequency, mainly from a point of view of phase noise and consumption, the values selected are shown in Table 8.1.

Once the signal of 3.2 GHz has been reduced to 800 MHz, we have the choice to employ another type of transistor (instead of the RF-optimized transistors used for frequencies above 1 GHz), which usually gives more flexibility regarding size and layout options.

In order to layout the differential to single-ended converter the following aspects must be taken into account:

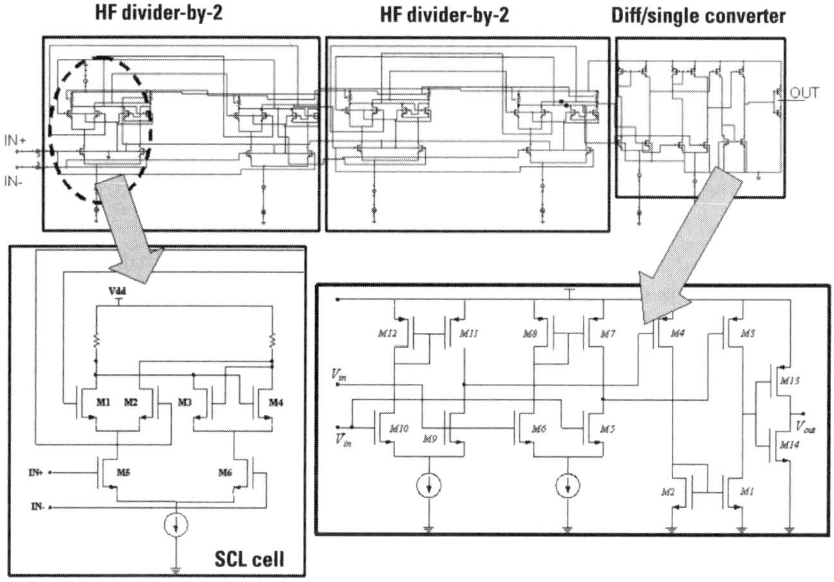

Figure 8.6 Schematic circuit of the two high frequency dividers-by-2 together with the differential to single-ended converter (first three blocks in Figure 8.1).

Table 8.1
Values of the Resistances and Bias Currents of Each Divider-by-2

	1° Divider	2° Divider
Current of polarization (mA)	1.7	0.5
Resistance (Ω)	500	1,200

- The input capacitance of the converter can affect the phase noise introduced by the second high-frequency divider-by-2.
- The rise and fall times of the output signal must be small enough to be stable, that is, it must not exceed the DC margins of the logic levels during the time of evaluation of the flip-flop of the following stage (low frequency divider-by-2).

By this method, with the objective of minimizing the input capacity of the converter, minimum dimensions for the transistors of the two NMOS differential amplifiers in Figure 8.2 (M5, M6, M9, and M10) must be chosen.

On the other hand, with reference to the two current sources of the converter, after carrying out different simulations, its value has been set at

50 μA. This value is negligible compared to the total consumption of the frequency divider but at the same time sufficient to generate the single-ended signal desired at the output of the converter.

Finally, we now present the design of the inverter. Due to the fact that the input frequency is still considerable, it is useful to employ a minimum length of channel for its transistors (0.18 μm). With this minimum length, it has been determined that the channel width of the PMOS transistor must be three times superior to that of the NMOS, due to their intrinsic mobility differences. With this assumption, we ensure the symmetry of high and low states and the noise margin [2]. Table 8.2 contains the values corresponding to the length and width of the channel of the transistors of this logic gate.

To finalize the design of these two initial blocks of the divider at schematic level, Figure 8.7 presents the input voltage together with the voltages at the output of each divider-by-2 and the converter obtained from the transient simulation of the circuit in Figure 8.6. It is important to point out that in order to obtain these results the output of the VCO in postlayout simulation has been used.

In the lower part of Figure 8.7, the input signal (3.2 GHz) can be seen at the first high frequency divider-by-2. Next, above this, the outputs of the first and second divider-by-2 and of the differential to single-ended converter are represented in sequence.

8.2.2 Low-Frequency Divider

As shown in Figures 8.3 and 8.4, this digital divider consists of various logic inverters and a single NAND gate. Given that the value of the module of this divider must be 40, it has been implemented using three dividers by 2 and a divider by 5 whose circuital diagrams have been demonstrated in Figure 8.6.

Due to the fact that the highest frequency is found at the input of the first divider-by-2, this has been implemented using minimum channel length transistors (0.18 μm). For the rest of the dividers, with the objective of reducing the power consumption, a channel length of 0.4 μm, has been chosen, due to two effects. The first one is that the stages with higher length transistors present

Table 8.2
Dimensions of the Transistors of the Inverter

	NMOS	PMOS
Length of channel (μm)	0.18	0.18
Width of channel (μm)	0.8	2.4

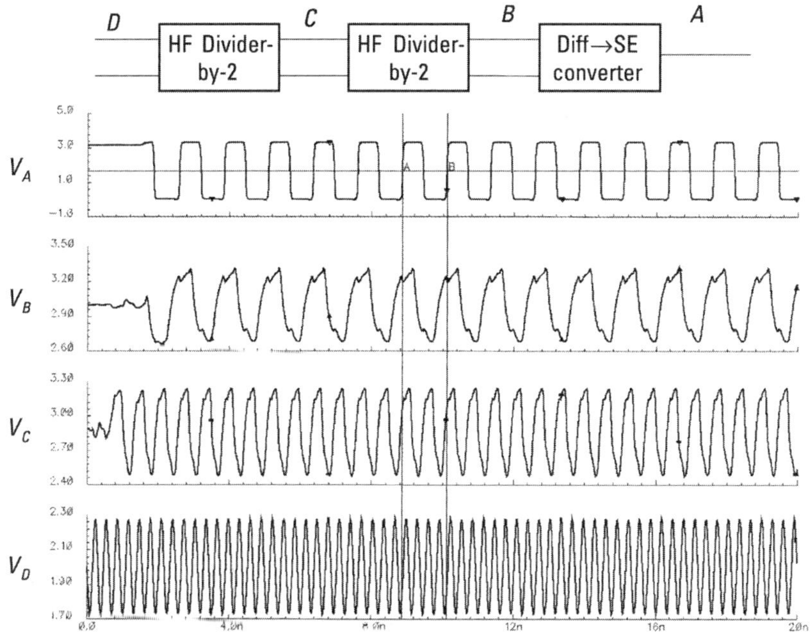

Figure 8.7 Results of the transitory simulation at schematic level of the two high-frequency dividers-by-2 together with the differential to single-ended converter.

higher impedance to the previous stages and no drivers are needed, and moreover, this reduces the variations owing to small channel effects.

With regards to the width of the channel, exactly the same thing occurs with the inverter of the single-ended to differential converter. With a length of channel of 0.18 μm or 0.4 μm, and after carrying out different simulations, it has been verified that the width of the PMOS transistors must be three times superior to that of the NMOS transistors. In the context of these considerations the values chosen for the NMOS and PMOS transistors of the logic gates of the low frequency digital divider are summarized in Table 8.3.

To finish this section the results of the transitory simulation of the complete digital divider are illustrated in Figure 8.8. The circuital diagram shown in Figure 8.9 has been used for this.

In the lower part of Figure 8.8 the input signal to the low frequency digital divider (800 MHz) can be seen (V_E). The output signals of the three dividers by 2 and the divider by 5 are situated above this (V_D, V_C, V_B). It can be observed that the period of the output signal from the complete digital divider is 50 ns (V_A), which represents a frequency of 20 MHz.

With the design of each one of the blocks of the divider now finalized, the global design of this component will now be presented.

Table 8.3
Dimensions of the Low-Frequency Divider Transistors

	NMOS	PMOS
Length of channel of the first divider-by-2 (μm)	0.18	0.18
Width of channel of the first divider-by-2 (μm)	0.8	2.4
Length of channel of the rest of the dividers (μm)	0.4	0.4
Width of channel of the rest of the dividers (μm)	0.8	2.4

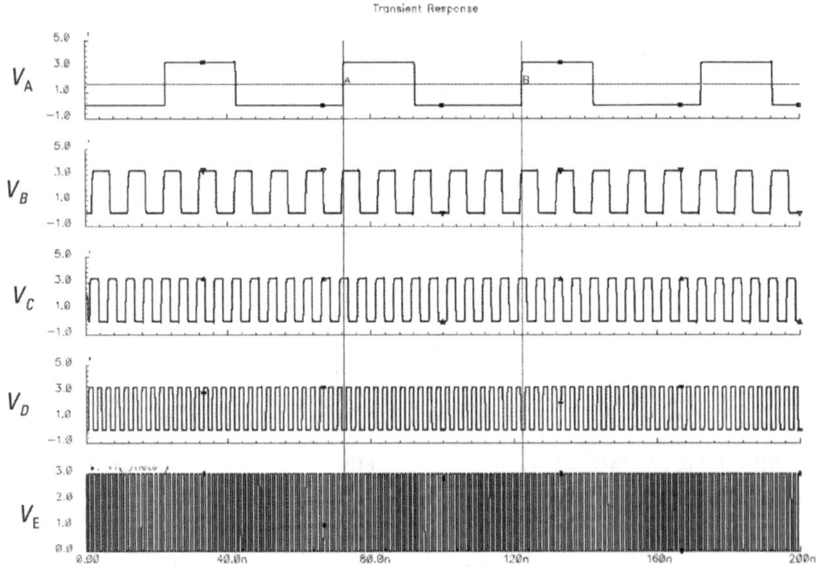

Figure 8.8 Results of the transitory simulation of the low frequency digital divider.

8.3 Design of the Schematic Circuit of the Divider

This section focuses on presenting the combination of the building blocks in order to complete its final schematic design. In order to carry out this design, the following steps have been taken:

1. Connection of the dividers of high- and low-frequency dividers previously presented and substitution of the ideal components by models supplied by the foundry. In this case the ideal components to be substituted are the current sources and resistors.

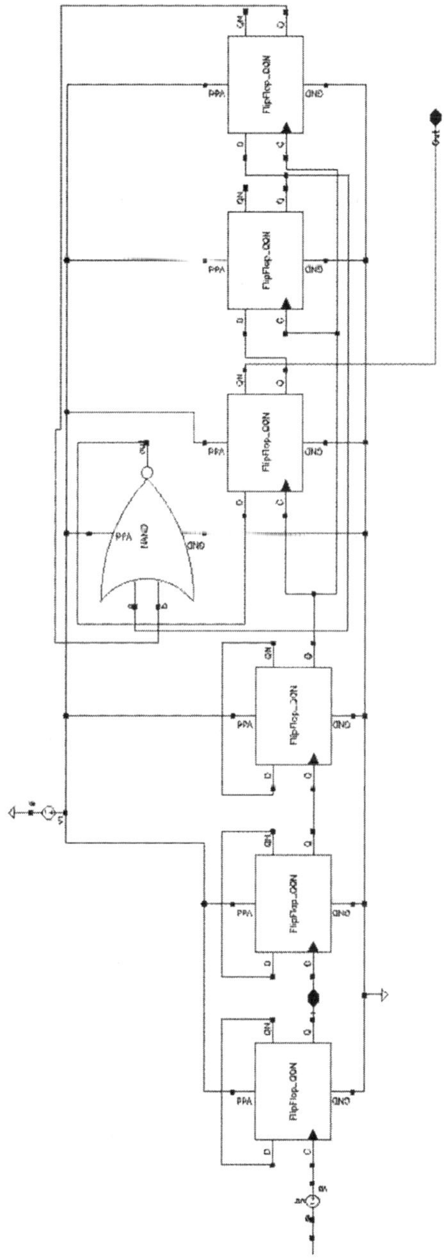

Figure 8.9 Schematic circuit of the complete digital divider.

2. Introduction of the elements required in order to obtain a more accurate simulation, that is, the output buffer, the connection pads, the decoupling capacitors between V_{DD} and ground, and the resistive voltage dividers.
3. Simulation of the final circuital diagram, leading to final selection of design parameters.

8.3.1 Connection of Building Blocks and Current Source Implementation

Once the blocks of the frequency divider have been joined together, the ideal resistors and current sources have been replaced according to their corresponding models.

As far as the current sources are concerned, the high-frequency divider and the single-ended to differential converter are the only two blocks that requires them. In both stages, current mirrors in cascode configuration have been used, in order to increase their output resistance. In addition to these current sources, the resistances of the high-frequency dividers-by-2 have also been substituted by their corresponding models.

8.3.2 Introduction of Auxiliary Components

The output buffer must face two different problems:

1. The output signal must track correctly the input, even with capacitive loads of various pF.
2. The circuit must be capable of supplying the necessary current to maintain logic levels of between 0V and 3.3V.

The two cases that have just been described can be found when the frequency divider is characterized through the use of an oscilloscope, whose input impedance is a resistance of 1 MΩ in parallel with a capacity of 12 pF, or a spectrum analyzer, whose input impedance is a resistance of 50Ω. Of course, the buffer should be adapted to the real working environment of the circuit, and may be removed in the fully integrated version of the circuit. However, it must be included for individual block characterization.

To design this output buffer, it must be taken into account that the response time of an inverter when it is loaded with a capacitance C_l can be obtained using:

$$t_{PHL} + t_{PLH} = (R_n + R_p) \cdot (C_{out} + C_l) \tag{8.1}$$

where t_{PHL} and t_{PLH} are the times that the signal takes to respond to the input, measured at mid-voltage (1.65V) of both signals, when the output signal commutes at a high logic level and the other low and vice versa, respectively. R_n and R_p are the effective resistances between the drain and source of the NMOS and PMOS transistors (obtained in both cases from the division V_{ds}/I_{ds}) and C_{out} is the output capacitance.

In the specific case of this example, the inverter placed immediately before the instrument of measurement (oscilloscope or spectrum analyzer) presents the same dimensions contained in Table 8.3. After estimating the values of R_n, R_p, and C_{out} of this gate and considering a load capacitance of 12 pF, the sum of the times t_{PHL} and t_{PLH} is 25 ns, which is an excessively high value compared to the period of the output signal from the divider (50 ns). Therefore, an alternative solution must be found.

If a buffer of inverters is introduced between this gate and the load, in a way that the channel widths of the transistors of each one are A times that of the previous, the sum of the times t_{PHL} and t_{PLH} of the chain is given in:

$$t_{PHL} + t_{PLH} = N \cdot (R_{n1} + R_{p1}) \cdot (C_{out1} + A \cdot C_{in1}) \tag{8.2}$$

where C_{in1} and C_{out1} are the input and output capacitances of the first inverter, R_{n1} and R_{p1} are the effective resistances between the drain and the source of the NMOS and PMOS transistors of the first inverter, N is the number of gates that form the buffer, and the value of A is obtained using:

$$A = \left[\frac{C_l}{C_{in1}}\right]^{1/N} \tag{8.3}$$

The number of inverters N_{min} that minimize the sum of t_{PHL} and t_{PLH} is obtained through the first derivative of (8.3) with respect to N. This result is given in:

$$N_{\min} = \ln \frac{C_l}{C_{in1}} \tag{8.4}$$

For the specific case of this work, the number needed is 7. By this method, after introducing the chain and estimating the values of C_{in1}, C_{out1}, R_{n1}, and R_{p1}, the value of the sums of the times t_{PHL} and t_{PLH} is obtained using (8.2) [2] to be approximately 0.5 ns, which is reasonable taking into account that the period of the output signal of the divider is 50 ns.

The schematic circuit of this buffer without including the initial inverter can be seen in Figure 8.10, and the scaling of the width of channel of each

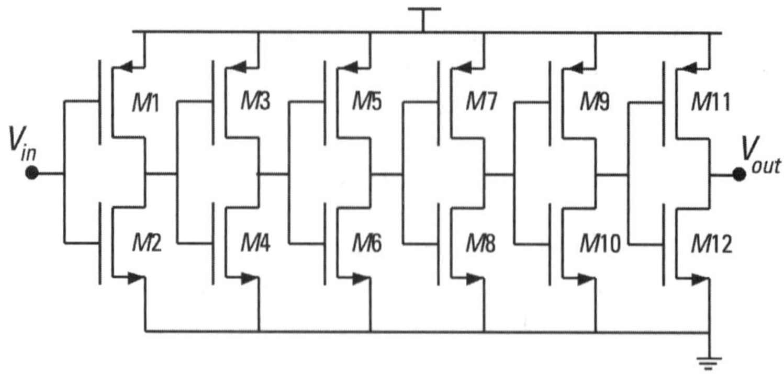

Figure 8.10 Output buffer of the divider.

one of its transistors are included in Table 8.4. It is important to point out that to minimize the sum of the times t_{PHL} and t_{PLH} it is enough to measure at intervals the width of channel, maintaining constant the length of this [2]. In this case, the channel length selected to implement the output buffer of the divider has a value of 0.4 μm. This is larger than 0.18 μm which is the length of the transistors of the core of the dividers in order to make these be connected to higher loads and then assure the level of the output amplitude. Moreover, small channel effects are avoided.

With reference to the connection pads, nine have been implemented: three RF input/outputs, two DC supplies and four grounds. Capacitors of high value (53 pF) between V_{DD} and ground voltages have been also included, in order to filter the noise coupled.

Finally, a resistive voltage divider has been placed in each one of the differential inputs in order to replicate the DC component of the output signals of the VCO output buffer. As seen in Section 7.2.3, this constant level is approximately 2V. This divider will be removed in the final integrated version of the PLL.

Table 8.4
Channel Width of the Transistors of the Buffer

	M1	**M2**	**M3**	**M4**	**M5**	**M6**
Channel Width (μm)	7	3	18	6	47	16
	M7	**M8**	**M9**	**M10**	**M11**	**M12**
Channel Width (μm)	126	42	340	113	910	304

After including all the components mentioned, the final schematic of the complete divider is presented in Figure 8.11.

The circuit shown in Figure 8.11 has been used to perform transient simulations, and the results can be seen in Figure 8.12. The period of the signal is 50 ns (20 MHz) and the levels of high and low voltages are 3.3V and 0V, respectively. Therefore, the rate of division of the complete divider at simulation of the circuital diagram level has a value of 160, which coincides with the value specified at the beginning.

In Figure 8.13 it can be established that the phase noise evaluated at an offset frequency of 1 kHz presents a value of −147.6 dBc/Hz. From this noise level it is easy to estimate the phase noise at the output of the PLL introduced by the frequency divider within the loop bandwidth, the issue developed in Section 4.6. Using the system transfer function, it is possible to carry out this estimation multiplying the noise by the squared division ratio. Therefore, the phase noise at the output of the PLL due to the frequency divider is approximately −103.51 dBc/Hz at an offset of 1 kHz. Comparing this result with that obtained at the same frequency in the initial simulation of the system (≈ −82 dBc/Hz), this can be considered to be negligible.

To conclude, it is interesting to note that the power consumption of the divider of the circuit is 19.2 mW. This does not include the output buffer, given that it is only necessary in order to obtain the experimental results of the component.

Table 8.5 summarizes the results obtained in the simulation of the divider.

8.4 Divisor Layout Generation and Simulation

The layout of the divider has been carried out following the recommendations presented in Section 4.7. The results of the postlayout simulation of this component will now be presented. These simulations have been carried out using the layout presented in Figure 8.14, where details of the high-frequency divider are illustrated in Figure 8.15.

As shown in Figure 8.14, the circuit has been implemented providing nine pads for external connections: the lateral left correspond to the differential divider inputs, with a central ground pad connection to ground. Regarding right pads, the one on the lowest is used to provide the supply voltage to the buffer. Finally, the SGS pads below are intended to bias the divider and to obtain the output signal.

The postlayout simulation of this component is presented in Figure 8.16. It can be seen again that the period of this signal is 50 ns, which means a frequency of 20 MHz. Therefore, it can be assumed that the module of the final divider in postlayout simulation is equal to 160.

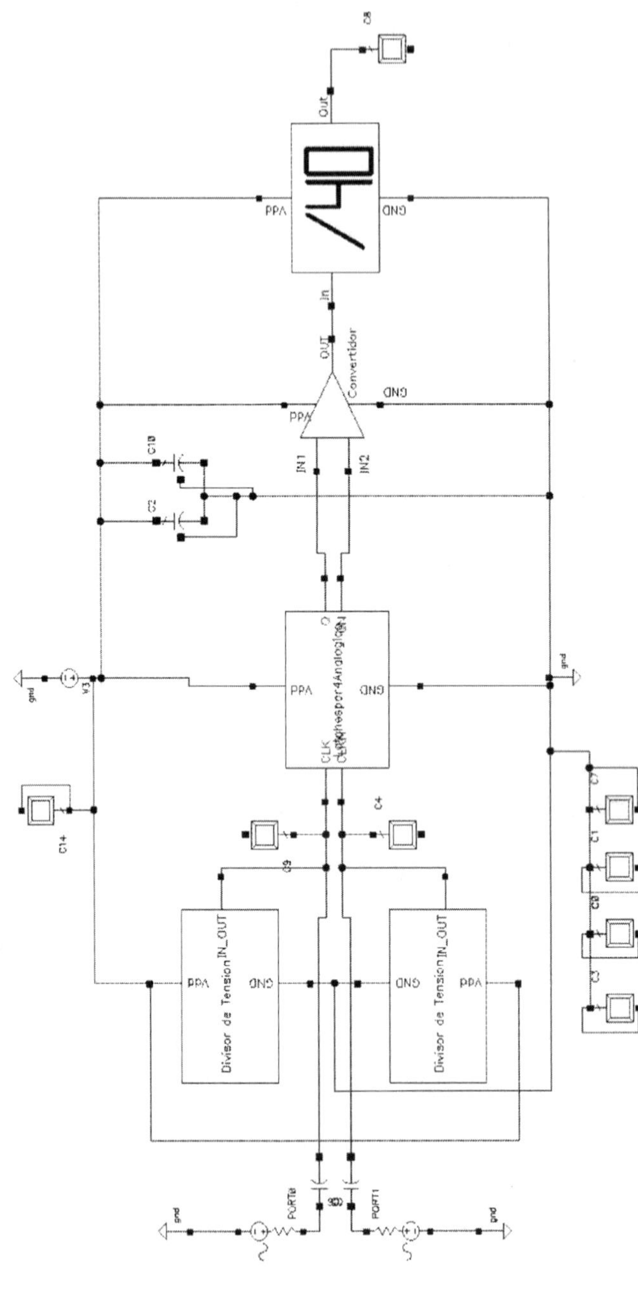

Figure 8.11 Complete circuital diagram of the frequency divider.

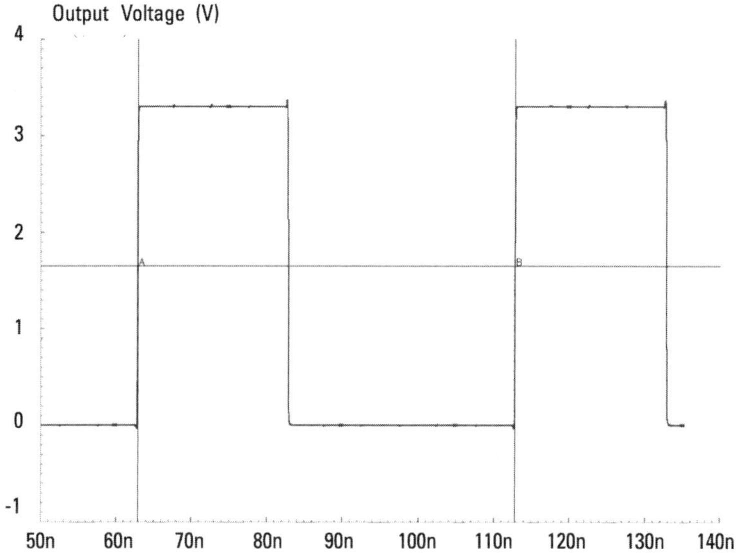

Figure 8.12 Output voltage versus time of the frequency divider.

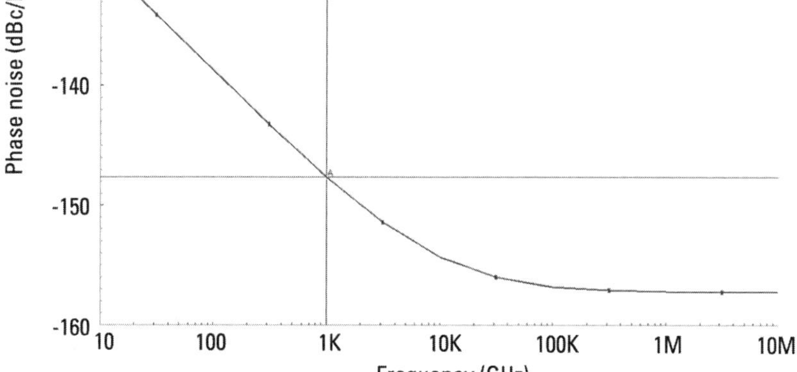

Figure 8.13 Simulated phase noise of the divider.

Table 8.5
Results of the Simulation of the Divider

Value of the module of division	160
Phase noise at 1 kHz (dBc/Hz)	–147.6
Consumption of power (mW)	19.2

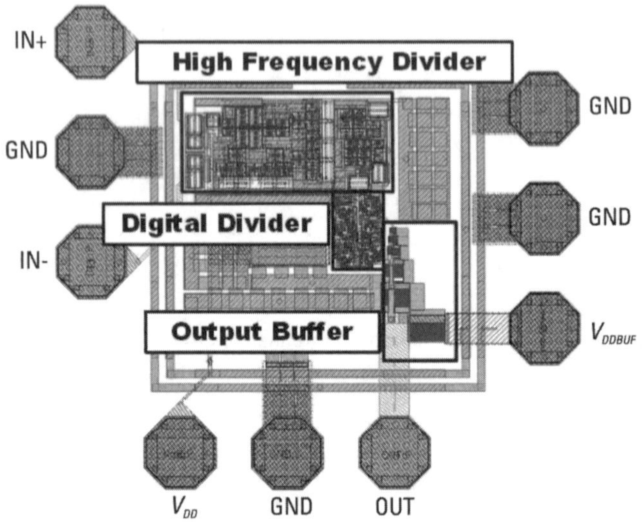

Figure 8.14 Layout of the complete frequency divider.

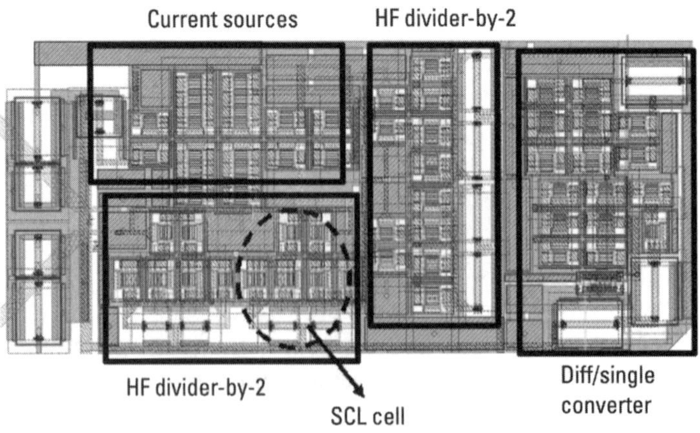

Figure 8.15 Layout of the high-frequency divider (schematic circuit shown in Figure 8.6).

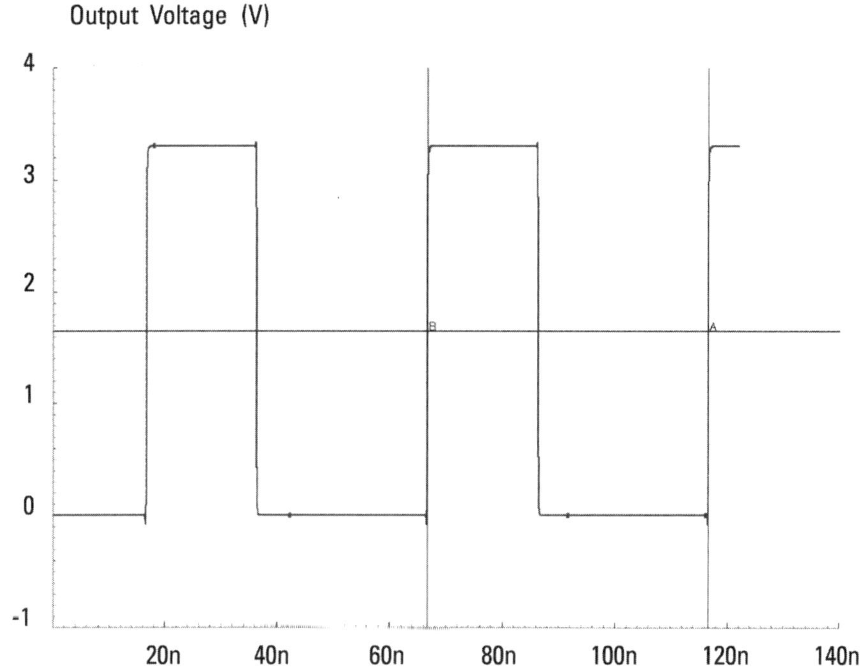

Figure 8.16 Output voltage from the divider obtained in the postlayout simulation.

The phase noise of the divider in postlayout simulation at an offset frequency of 1 kHz presents a value of −145 dBc/Hz, which indicates it can be neglected compared to other phase noise contributions.

The total power consumption of the divider obtained in the postlayout simulation hardly varies with respect to the result of the schematic circuit, presenting a value of 19.1 mW.

Table 8.6 contains the results of this postlayout simulation.

Table 8.6
Results of the Postlayout Simulation of the Divider

Value of the module of division	160
Phase noise at 1 kHz (dBc/Hz)	−145
Power consumption (mW)	19.1

References

[1] Maneatis, J. G., "Low-Jitter Process-Independent DLL and PLL Based on Self-Biased Techniques," *IEEE Journal of Solid-State Circuits*, Vol. 31, No. 11, November 1996, pp. 1723–1732.

[2] Baker, R. J., H. W. Li, and D. E. Boyce, *Circuit Design, Layout, and Simulation*, New York: IEEE Press, 1998.

9

Design of a Phase Frequency Detector

After the presentation of the implementation of the oscillator and the frequency divider, this chapter describes the steps followed in the design of the phase frequency detector. First, as with the two previous blocks, the choice of the most suitable architecture among the PFD/PD alternatives will be justified, followed by a description of the design steps followed.

The starting specifications of the phase frequency detector that must be fulfilled and that have been determined in Chapter 6 are contained in Table 9.1.

9.1 Choice of the Architecture of the Detector

In Chapter 5 the types of phase frequency detector/phase detectors most commonly used have been described. These are classified in four categories:

- Analog or multiplier-based phase detectors;
- Basic sequential circuits, as logic gate OR-exclusive;
- Sequential flip-flop phase detector;

Table 9.1
Specifications of the Phase Detector

K_ϕ (μA/radian)	L_{1Hz} (dBc/Hz)
50	<-200

- Sequential circuits with the extra characteristic of supplying a signal sensitive to the frequency that helps lock the loop when the PLL is unlocked. The most used phase detector of this type is the PFD/CP.

The first category of phase detectors has been discarded for this work as a consequence of the high number of unwanted frequency components generated. Among the three categories left, the PFD/CP has been selected for the following reasons:

- It acts as a phase detector while the loop remains locked and it supplies a signal sensitive to the difference of frequency between its inputs that helps in reaching the locked phase when the PLL is unlocked.
- It has a linear range of input (±360°) superior to that of the logic gate OR-exclusive and the flip-flop, and in addition, it is sensitive to the sign of the phase error between its inputs.
- It is suitable for integrated realizations.

The configuration selected to implement the PFD/CP phase detector is presented in Figure 9.1, where the reset path block has the main objective of

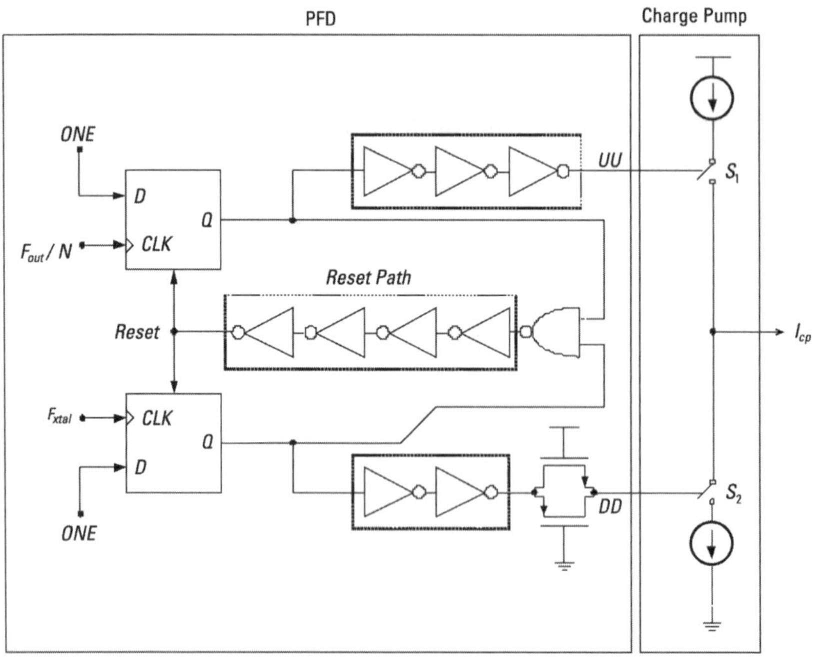

Figure 9.1 Architecture of the phase detector selected.

eliminating the dead zone of the detector. The calculation of the number of inverters necessary to achieve this objective will be presented in the section dedicated to the design of this component.

As can be seen in Figure 9.1, a delay gate has been introduced in the lower branch of the PFD in order to eliminate the possible time difference between *UU* and *DD* as a consequence of using an extra inverter in the upper branch due to the signal polarity requirement. It is necessary to use this extra inverter, as will be described next, due to the fact that the current source of the charge path is implemented with PMOS transistors, unlike that of the discharge that is developed with NMOS. This small dephase could influence the level of the reference spurs, as described in Chapter 5. The schematic circuit selected for each of the PFD flip-flops is presented in Figure 9.2, where *NAND*1 and *NAND*2 are two NAND gates. The first is conventional, while the second is controlled by two signals with 180° out of phase *CLK* and $\overline{CLK}$. When the reset input of the flip-flop (*R*) is at a low level, the output of *NAND*2 will always be at a high level (the reset of the flip-flop is activated). On the other hand, when the reset input is at a high level (the flip-flop functions normally, the output of *NAND*2 inverts with respect to the input *b* only if the *CLK* input is also at a high level. Finally, the diagrams for *IN*1 and *IN*2 are the same as those used in the low frequency digital divider flip-flops (as seen in Figures 8.4 and 8.5).

Therefore, the configurations of the logic gates used to implement the diagram of the PFD in Figure 9.1 are the same as those used in the low frequency digital divider except the topology of the gate *NAND*2, whose schematic circuit design appears in Figure 9.3.

To conclude, the architecture chosen at transistor level to implement the PFD charge pump is illustrated in Figure 9.4.

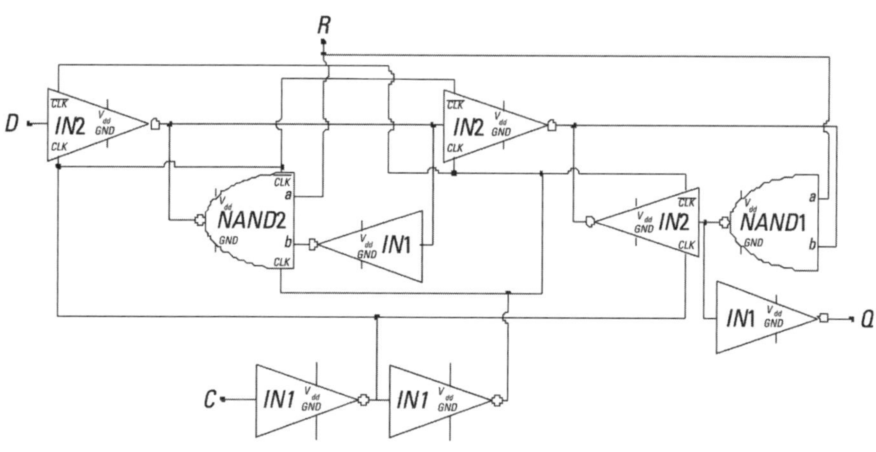

Figure 9.2 Configuration of logic gates of each PFD type D flip-flop.

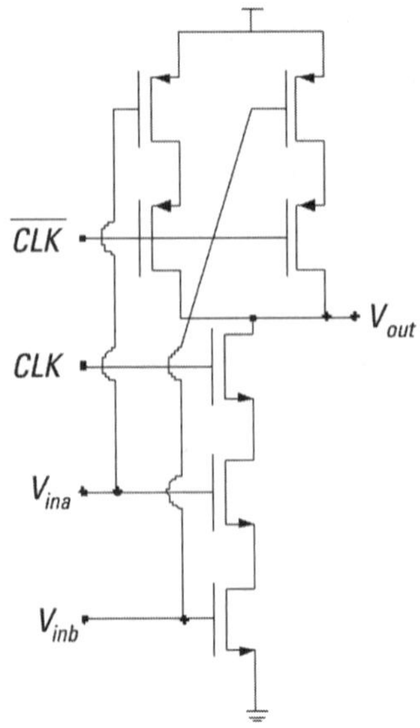

Figure 9.3 Schematic circuit design of the logic gate *NAND*2.

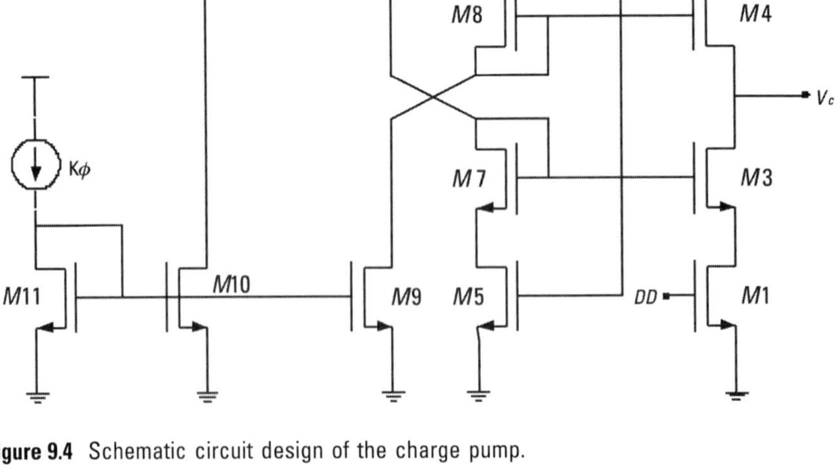

Figure 9.4 Schematic circuit design of the charge pump.

The following are the reasons that justify the choice of this architecture:

- In this topology unlike the conventional architecture, $M1$ and $M2$ act as switches instead of $M3$ and $M4$. This mitigates the errors of charge injection and clock feedthrough at the output of the charge pump and as a consequence reduces the jitter in the voltage control of the VCO that gives rise to reference spurs.
- Usually, the low output impedance of $M3$ and $M4$ gives rise to important mismatches between the charge and discharge currents of the CP, especially in the extremes of voltage control (V_c). In the specific case of this design, the transistors that act as *switches* $M1$ and $M2$ raise this impedance reducing the mismatch of current and in short the jitter in the voltage control of the oscillator.
- It is a simple architecture which is easy to integrate.

Having justified the choice of the architecture of the phase detector, the phase of design of the component will now be presented.

9.2 Design of the Phase Frequency Detector

The objective of this section is to present the implementation of the phase frequency detector, from the design to the schematic level until the complete development of its layout.

9.2.1 Design of the PFD

First, the implementation of the PFD will be explained. The most important aspect of the design of this block is without doubt the estimation of the delay of the reset path. In Chapter 5 it was stated that one of the fundamental disadvantages of the PFD/CP topology is the dead zone that appears in its transfer characteristic. This dead zone represents the difference of phase between the inputs of the PFD for which the charge pump is found to be inactive. This effect is produced by small differences of phase between its inputs given that when this happens the output pulses of the PFD become so narrow that they are not able to activate the charge pump.

Therefore, in order to eliminate the dead zone, it is interesting to increase the width of these pulses until they reach a valid logic level and are therefore able to activate the charge pump. In order to achieve this it is necessary to introduce a delay in the reset path of sufficient value in order to obtain this effect.

The minimum delay necessary that must be introduced in the reset path has been estimated as the mean value of the rise and fall times of the pulses of the nodes of output *UU* and *DD* [1]. This delay has been then compensated by means of four inverters. Logically the number of gates must always be even in order to not reverse the global logic state.

Now that the calculation of the number of inverters necessary in the *reset* path has been described, the selection of the dimensions of the logic gates that comprise the PFD will be justified. Given that the working frequency is not excessively high (20 MHz), a channel length of 0.4 μm has been chosen for all its transistors. Nonminimum transistors have been selected, even if higher capacitances are added to this reset path, because the delay needed is achieved with a lower number of stages. Using the same reasoning as described in the design of the low frequency digital divider (Section 8.2.2), for the PMOS transistors a width three times larger has been selected. Table 9.2 contains the values of the channel length and width of the transistors corresponding to the PFD logic gates.

With regards to the delay element necessary in order to eliminate the dephase existing between signals in *UU* and *DD* nodes, the length and width of its transistors have been selected in a way that the delay is cancelled in the middle of rising pulses of both signals, as shown in Table 9.3.

The final schematic circuit design of the PFD, after introducing the length and width of the channel of the transistors of each one of its logic gates can be seen in Figure 9.5.

From this schematic design and performing a transient simulation, the output signals in the nodes *UU* and *DD* are presented in Figure 9.6, when the delay between the input signals to the PFD is ±10 ns.

Table 9.2
Transistor Dimensions of the PFD Logic Gates

	NMOS	PMOS
Channel length (μm)	0.4	0.4
Channel width (μm)	0.8	2.4

Table 9.3
Dimensions of the Transistors of the Delay Element

	NMOS	PMOS
Channel length (μm)	0.4	0.4
Channel width (μm)	2.4	2.4

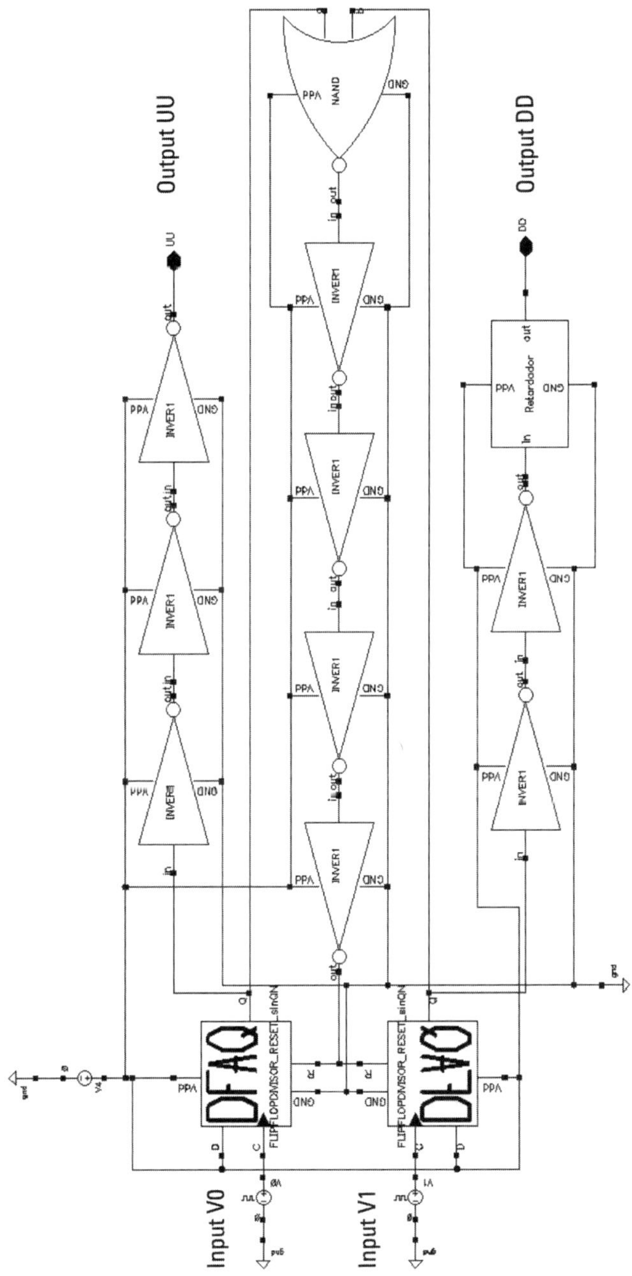

Figure 9.5 Schematic circuit design of the PFD.

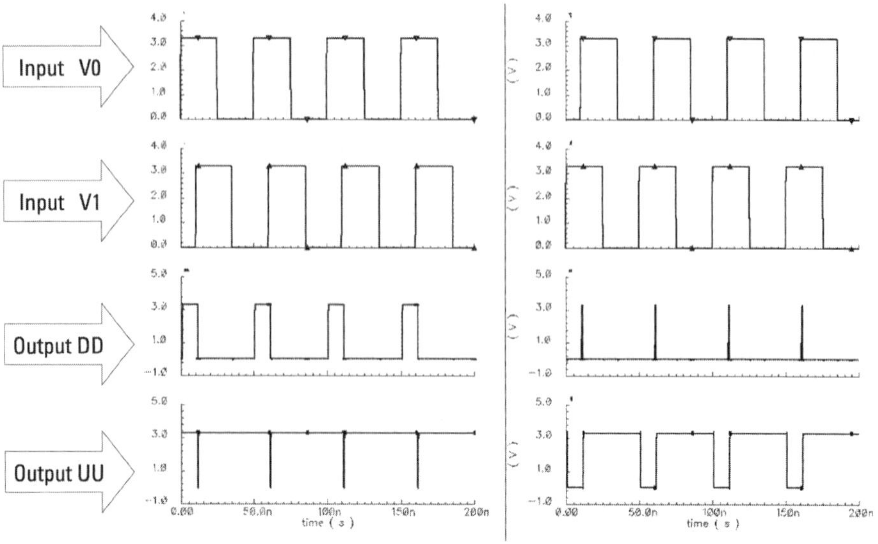

Figure 9.6 Output signals of the PFD for an input phase difference of ±10 ns.

In Figure 9.6 the output signals of the PFD can be seen (outputs *UU* and *DD* of Figure 9.5). Depending on the phase difference between its input signals (two upper rows of plots in Figure 9.6) and which of them is more advanced with respect to the other, the PFD generates pulses through one of its inputs of a width proportional to this phase difference (two lower rows of plots in Figure 9.6). It is important to point out that if the signal of the reference crystal is advanced with respect to that proceeding from the frequency divider, the PFD generates pulses through its output *DD*. On the contrary, the pulses are generated through the output *UU*, which are also of low logic level in order to be able to commute the PMOS transistors of the subsequent current source.

Next, the design corresponding to the second stage of the detector, the charge pump, is described.

9.2.2 Design of the Charge Pump

This has great influence on the reference spurs at the output of the loop, which is why it is necessary to take into account certain considerations in order to try to reduce its impact.

The diagram proposed for the charge pump has already been presented in Figure 9.4. As can be seen in this figure the critical dimension corresponds to transistors $M1$, $M2$, $M3$, and $M4$, that govern the switching of the charge pump, while the rest of current mirrors are necessary to copy the value of specified current (50 μA) at the output. The size of the four previous transistors, which coincide respectively with those of *M*5, *M*6, *M*7, and *M*8 to mirror the current as accurately as possible, must be carried out taking into account principally the speed of commutation of the NMOS and PMOS switches.

As explained in Section 5.6.2 and as shown in Figure 9.4, it can be seen that the transistors that act as switches (*M*1 and *M*2) vary between the regions of operation cutoff and triode, while *M*3 and *M*4 oscillate between cutoff and saturation. Two resistances are in series in each branch. From all these considerations it is now possible to present the process followed in the selection of the dimensions of the transistors that perform the commutation in the charge pump, following the guidelines provided in Section 5.6.2.

In this process identical dimensions for *M*1 and *M*2 have been chosen so that they are switching correctly at the frequency of operation of the charge pump (20 MHz), while the sizes *M*3 and *M*4 ensure that the switching speeds of both branches are the same. In this way, as the NMOS transistors are faster than those of PMOS, it is necessary for the *W*/*L* ratio of *M*4 to be larger to that of *M*3 in order to reduce the total channel resistance of the upper branch. In addition to all this, it needs to be taken into account that the parasitic capacitance introduced will vary depending on the dimensions of the selected transistors. As mentioned previously, the speed of commutation decreases in accordance with the increase in C_H. The situation can arise whereby in attempting to increase the speed of commutation by an increase of *W*/*L*, this effect is compensated by an increase of C_H up to the point where no improvement in speed can be noticed.

Due to the high number of variables and effects involved, it is recommended to use the circuit simulator in order to obtain a suitable combination of all parameters. The target of the optimization process is to obtain no phase difference between the current generated in both branches of the charge pump.

Figure 9.7 shows the circuit used to perform this optimization, obtained from a combination of circuits previously shown in Figures 9.4 and 9.5, while Figure 9.8 shows the absolute value of output currents. The maximum difference between both charge and discharge currents is less than 2.3 μA, something which means a maximum mismatch between them is less than 6%, which is acceptable for the correct PLL behavior.

On the basis of these premises, the final values selected for the length and width of the channel of the transistors of the charge pump are demonstrated in Table 9.4.

With the charge pump completed, it is now possible to combine the design of the two blocks in order to obtain the final results at schematic level of the phase detector.

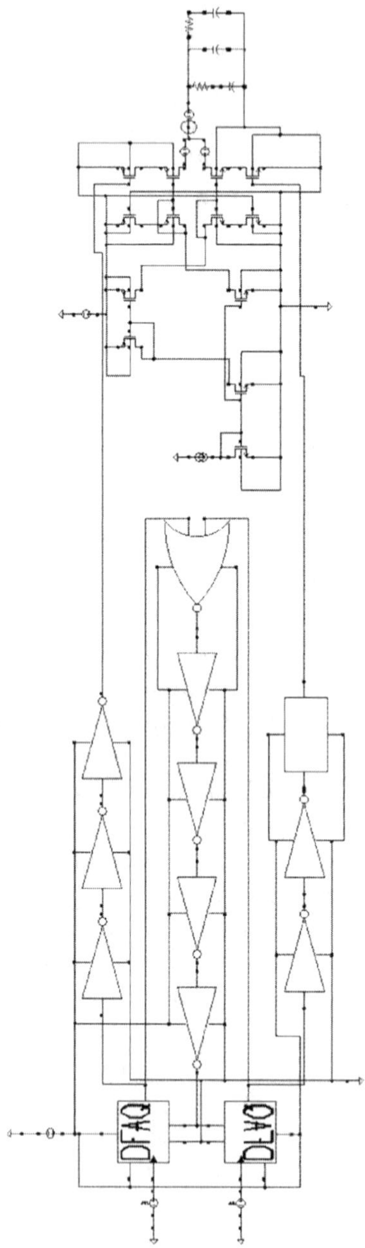

Figure 9.7 Schematic circuit design of the charge pump together with the PFD (from Figures 9.4 and 9.5).

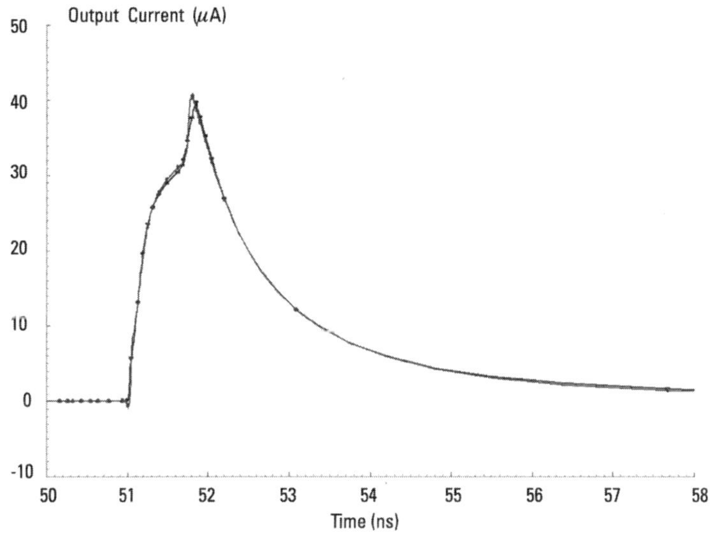

Figure 9.8 Absolute value of the charge and discharge output currents.

Table 9.4
Dimensions of the Transistors of the Charge Pump

	*M*1, *M*2, *M*5, *M*6	*M*3, *M*7	*M*4, *M*8	*M*9
Channel length (μm)	0.4	4	2.4	4
Channel width (μm)	3.5	5	4	6
	***M*10**	***M*11**	***M*12**	***M*13**
Channel length (μm)	4	4	3	3
Channel width (μm)	6	6	9	9

9.2.3 Design of the Schematic Circuit of the Phase Frequency Detector

Once the different blocks of the phase detector have been successfully combined, the ideal current source of the charge pump has been substituted by a current mirror in cascode configuration. Following this, the models of the connection pads and the decoupling capacitors of V_{DD} are introduced with the same objective as that described in the design of the previous blocks of the frequency synthesizer.

After introducing all these elements, the final schematic circuit design can be seen in Figure 9.9. The simulations have been performed using two square signal generators with variable phase difference, which account for the reference crystal and the output of the frequency divider.

In order to obtain the nominal value of the charge and discharge current of the charge pump, a transient simulation of the circuit has been carried out for a phase difference between the inputs of ±25 ns. These input signals, together with the output current of the charge pump for the two differences of phase, are presented in Figure 9.10. It can be seen that depending on the sign of phase difference, the charge pump acts like a source or drain of current of a value close to 50 μA/radian.

Once the values of the charge and discharge currents have been obtained, the noise $L_{1\text{Hz}}$ from the phase detector can be calculated.

It has already been explained in detail in Section 5.5 that the noise from the phase detector normalized at the reference frequency of 1 Hz is an eminently empirical parameter. However, in [2] a method of calculation is proposed. This alternative method is a practical approach very useful for any case studied.

It is based on the temporary jitter that appears in the pulses of current of the CP when the PLL is locked. This jitter is represented by a variable t_k of average value equal to zero and variance $\sigma^2 = \overline{t_k^2}$. The noise $L_{1\text{Hz}}$ is calculated from this jitter using (9.1).

$$L_{1\text{Hz}} = 20 \cdot \log(2 \cdot \pi \cdot \sigma) \tag{9.1}$$

In order to be able to calculate the variance of the jitter at the input of the charge pump using the simulator it is necessary to perform specific noise simulation of periodic signals. By means of this analysis, the simulator permits the calculation of $\overline{n(t, \omega)^2}$, which represents the contribution of all the sources of noise from the phase detector at the frequency of noise ω for each instant of time of the period.

After carrying out this analysis assuming that the PLL is locked (phase difference equal to zero) the spectral density of noise in $A/\sqrt{\text{Hz}}$ for each instant of time of the period together with the average value of this can be seen in Figure 9.11.

According to the study carried out in [2], this spectral density of noise in A^2/Hz can be obtained using (9.2).

$$\frac{\overline{i^2}}{\Delta f} = \sigma^2 \cdot F_{ref} \cdot K_\phi^2 \tag{9.2}$$

Substituting the mean value of the spectral density of noise given by the simulator, the reference frequency F_{ref} and the value of the current of the charge pump

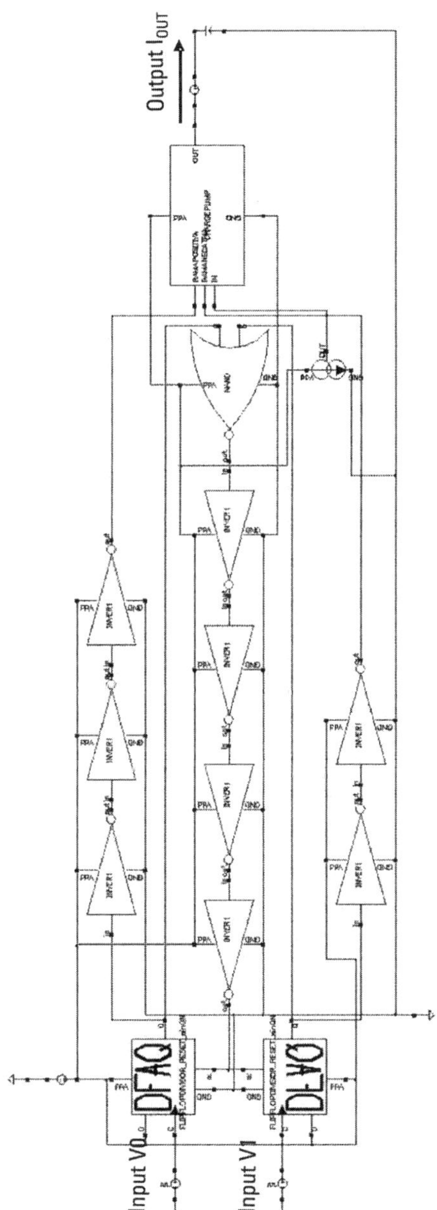

Figure 9.9 Schematic circuit design of the PFD/CP.

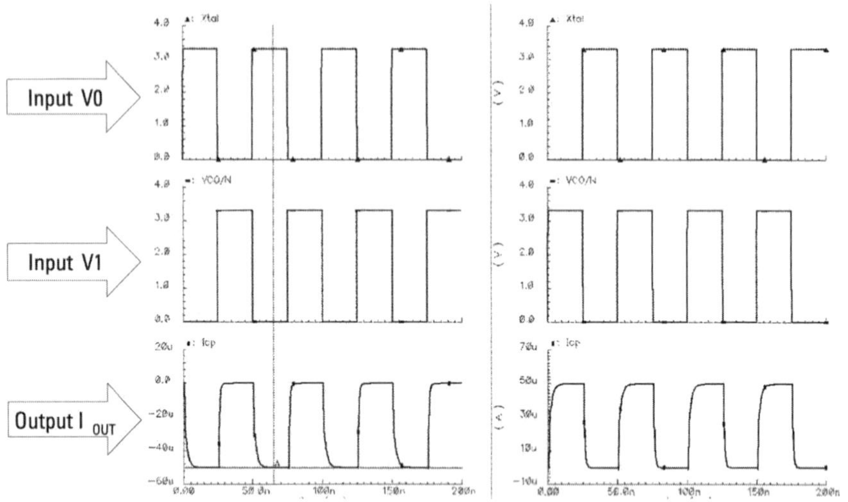

Figure 9.10 Output current of the CP for an input phase difference of ±25 ns.

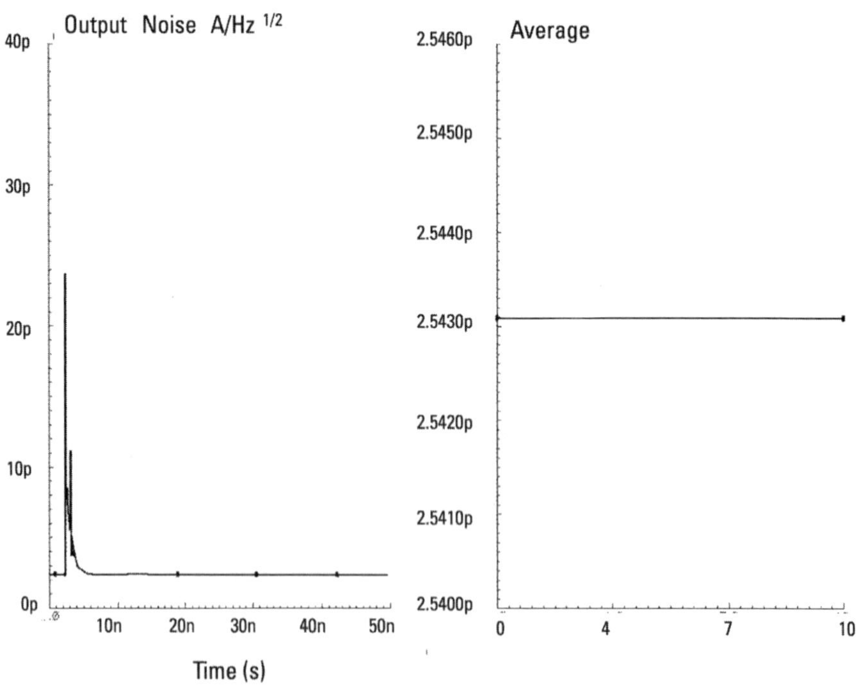

Figure 9.11 Density of current of noise from the phase detector at schematic level.

K_ϕ, it is found that σ is approximately 11.3 ps. From this value and using (9.1), the noise $L_{1\text{Hz}}$ from the phase detector is estimated at −202.92 dBc/Hz, which is approximately 3 dB less than the initial specification contained in Table 9.1.

Table 9.5 proves that the results corresponding to the value of the current of the charge pump and to the noise $L_{1\text{Hz}}$ fulfill the specifications set out in Table 9.1. Therefore, with the schematic circuit design of the phase detector finalized, it now becomes possible to begin the implementation of the layout of this component.

9.2.4 Postlayout Simulations of the Phase Detector

As with the previous devices, the last phase of the design of the phase detector consists of implementing its layout and verifying that the results obtained in the postlayout simulation also fulfill the initial specifications of this component. The layout implemented is shown in Figure 9.12, which has been defined following the rules explained in Section 5.3.

In order to check the consistency of this layout, postlayout simulations, including all the parasitic effects, have been carried out. The results show an absolute value of charge and discharge current of approximately 50 μA, similar to the schematic simulation.

The same simulation as that described previously has been carried out for the calculation of the noise $L_{1\text{Hz}}$. The spectral density of the noise in $A/\sqrt{\text{Hz}}$ for each instant of time of the period, together with the average value of this, can be seen in Figure 9.13.

Using the mean value of the spectral density of noise in Figure 9.14 and from (9.1) and (9.2), it is found that the value of $L_{1\text{Hz}}$ in postlayout simulation equals −202.31 dBc/Hz. The third result of interest from the postlayout simulation is the total power consumption of the PFD/CP, which reaches the same value of 0.5 mW as in the schematic simulation.

Finally, it is interesting to check the correct behavior of the circuit implemented to cancel the dead zone. For this purpose, the mean value in a period of the output current of the charge pump for a phase difference between the inputs of ±5 ns is presented in Figure 9.14.

Table 9.5
Final Results of the Phase Detector at Schematic Circuit Design Level

K_ϕ (μA/radian)	$L_{1\text{Hz}}$ (dBc/Hz)	Consumption (mW)
50	−202.92	0.5

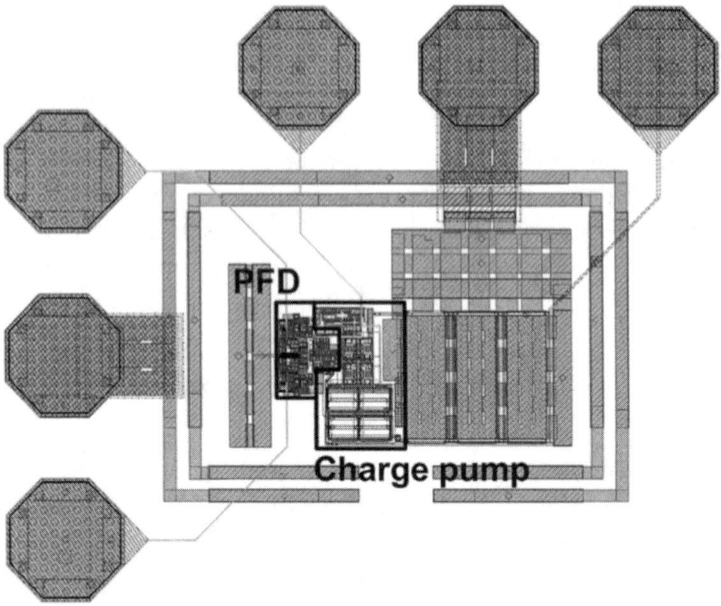

Figure 9.12 Layout of the phase detector.

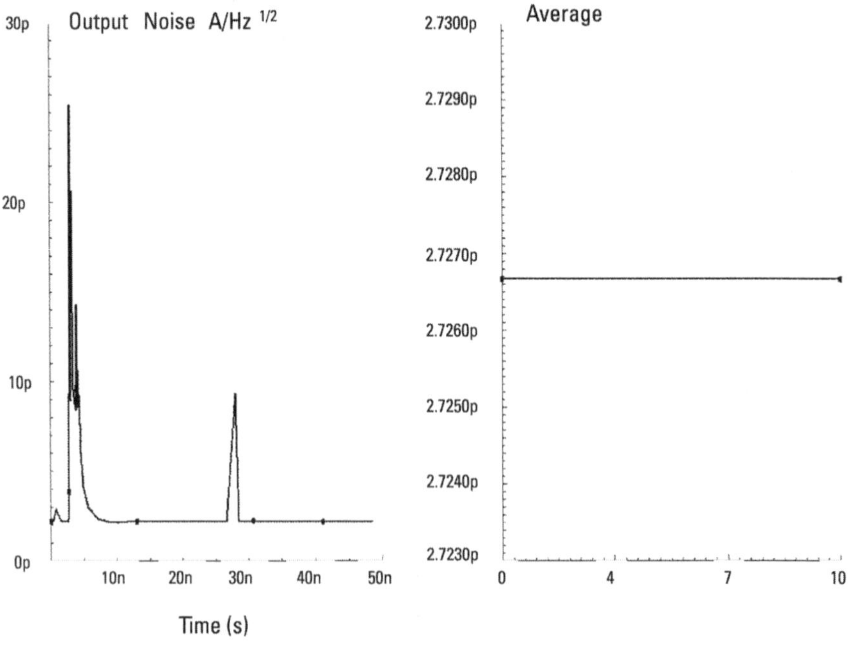

Figure 9.13 Density of current of noise from the phase detector at postlayout level.

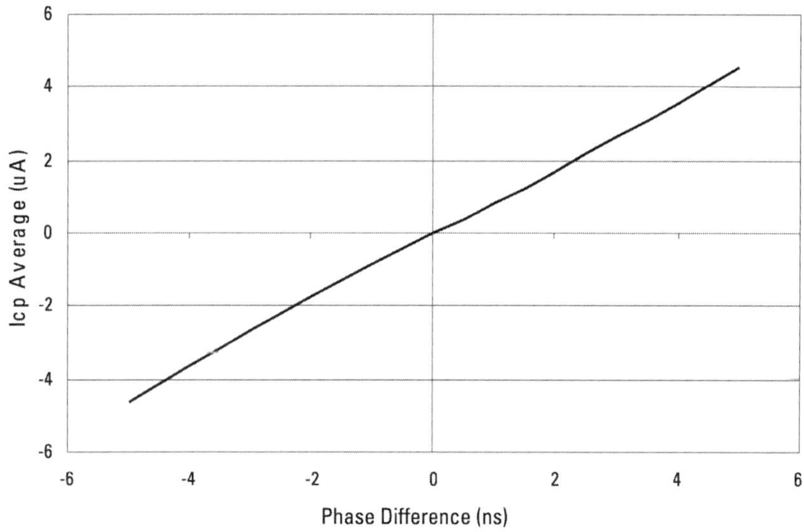

Figure 9.14 Transfer function of the phase detector in postlayout simulation.

It can be seen in Figure 9.14 that the dead zone of the phase detector has been cancelled thanks to the correct estimation of the delay introduced by the reset path of the PFD. Finally, the summary of the specific values obtained at postlayout simulations of the phase detector are contained in Table 9.6. It can be observed that they fulfill all the initial specifications required.

Table 9.6
Comparison Between the PFD/CP Specifications and Simulated Values

	Specified	**Simulated**
K_ϕ (μA/radian)	50	50
L_{1Hz} (dBc/Hz)	–200	–202.31

References

[1] Lee, H., et al., "Scheme for No Dead Zone, Fast PFD Design," *Journal of the Korean Physical Society*, Vol. 40, No. 4, April 2002, pp. 543–545.

[2] White, P., "Understanding Phase Noise from Digital Components in PLL Frequency Synthesizers," Applied Radio Labs, December 2000.

10

Design of the Complete PLL

In the previous chapters the design of the three principal blocks of the frequency synthesizer has been presented. On the basis of these designs, this chapter explains the development and implementation of the complete locked loop together with the presentation of its most important characteristics.

First, the starting requirements of the synthesizer are demonstrated. Later the design of the frequency synthesizer from its schematic circuit design to the implementation of the definitive layout is explained. This point of design is especially important, given that from the results obtained from Simusyn and from the electrical circuit simulation tool it is verified whether the final results of the PLL, before its manufacture, meet the initial requirements.

10.1 General Considerations

The block diagram of the architecture of the synthesizer selected in Chapter 5 is presented in Figure 10.1.

Table 10.1 summarizes the requirements of the frequency synthesizer presented in this book, determined in Section 6.1 together with the minimum amplitude required at the output of the LC-tank, specified in Section 6.4.2.

It is important to point out that the requirement corresponding to the lock time of the synthesizer has not been included in Table 10.1 given that it does not constitute a real specification of this work. However, it is possible to estimate its value by means of Simusyn, assuming a maximum frequency jump of 200 MHz.

Having summarized the requirements of the synthesizer that must be fulfilled, the design of this component will now be presented.

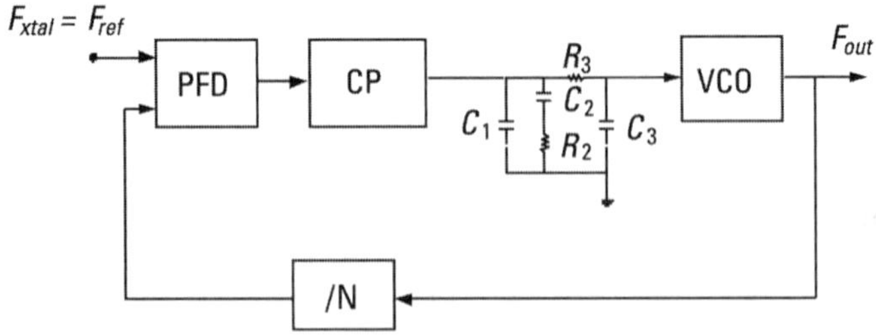

Figure 10.1 Architecture of the frequency synthesizer.

Table 10.1
Initial Requirements of the Synthesizer

Output Frequency (GHz)	3.2
Reference Frequency (MHz)	20
Reference Accuracy (ppm)	±20
Phase noise at 1 MHz (dBc/Hz)	–110
Integral of the total phase noise over BW_{canal} (dBc)	–32
Reference Spurious Level (dBc)	–59
LC-tank differential amplitude (V)	>0.5

10.2 Schematic Circuit Design of the Synthesizer

As with each one of the blocks of the frequency synthesizer, in order to obtain the design of the schematic circuit of the PLL the following steps have been taken:

- Combination of each block of the loop presented in Chapters 7, 8, and 9 together with the filter of the loop, calculated using the Simusyn tool;
- Introduction of the elements necessary to obtain a more accurate schematic simulation, that is, connection *pads* and additional decoupling capacitors;
- Simulation of the total schematic circuit followed by the extraction of results.

Following these steps, the final schematic circuit design illustrated in Figure 10.2 has been obtained. It is important to point out that the components of the loop filter have been calculated by means of the Simusyn tool whose values, together with the inputs of this tool necessary to calculate them, are contained in Table 10.2. As it can be observed, the high value of these components would

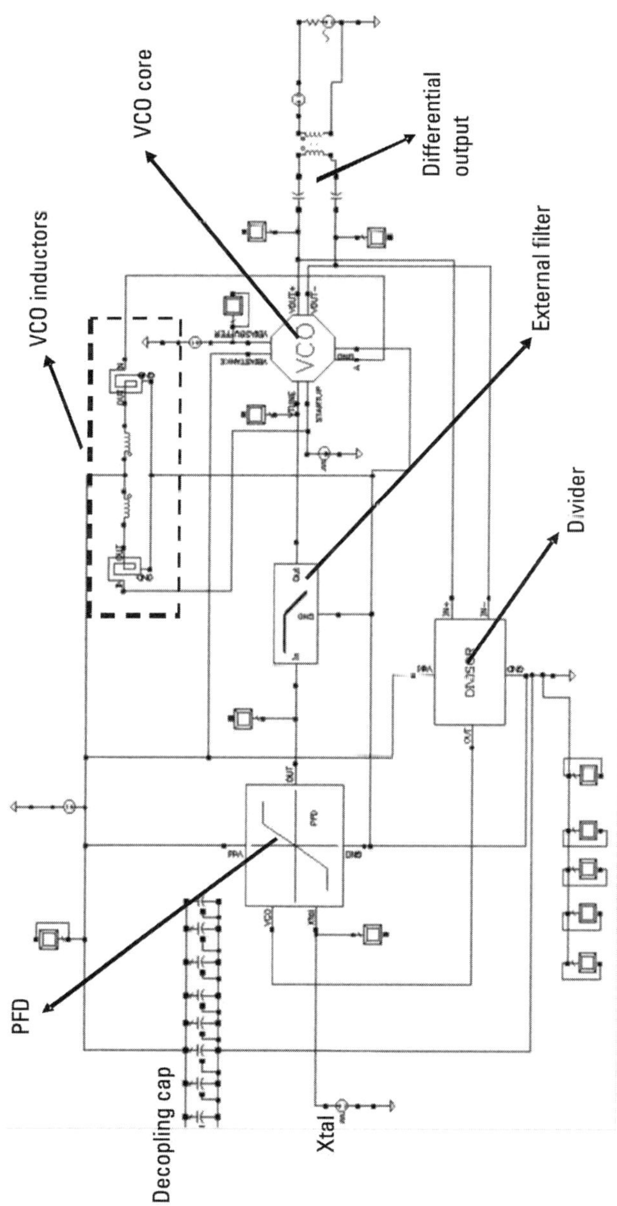

Figure 10.2 Complete circuit design of the frequency synthesizer

Table 10.2
Values of the Components of the Filter of the Loop

Parameter	Value
N	160
R	1
F_{xtal}	20 MHz
$ATTEN$	20 dB
K_{VCO}	170 MHz/V
ϕ_p	55°
f_c	20 MHz
F_{out}	3.2 GHz
K_ϕ	50 μA
$C1$	780 pF
$C2$	3.4 nF
$C3$	78 pF
$R2$	3.9 KΩ
$R3$	306 Ω

lead to an extreme area occupation in an integrated realization, therefore the use of an external filter constitutes the most logical and suitable option.

In order to increase accuracy of these simulations, the gain of the VCO (K_{VCO}, which is obtained from the ratio of the maximum bandwidth range over the variable control voltage) used at this point has been obtained in the postlayout simulation of the oscillator. It is consequently closer to its final real value, for the sake of a greater exactitude in the estimation.

From the schematic circuit design in Figure 10.2, it is possible to obtain the first simulated results of the frequency synthesizer. The results that are presented throughout this section are the output frequency, the phase noise and its integral over the bandwidth of the channel (20 MHz), the level of reference spurious emissions, the differential amplitude at the output of the VCO tank and the lock time of the PLL.

To calculate the output frequency of the PLL and the level of spurious signals, a transient simulation has been performed. With this it is possible to verify the correct evolution of the loop. The total level of reference spurs were determined carrying out a *discrete Fourier transform* (DFT) at the output signal of the synthesizer. The result of this DFT is shown in Figure 10.3 and is clearly more instructive than the quasi-perfect transient sine wave of the locked output. Figure 10.3 offers specific values of the main frequency tone and of the spurious signals.

In this figure it can be seen how the synthesizer is found to be locked at the frequency of output of 3.2 GHz with an output power of 2.9 dBm and a

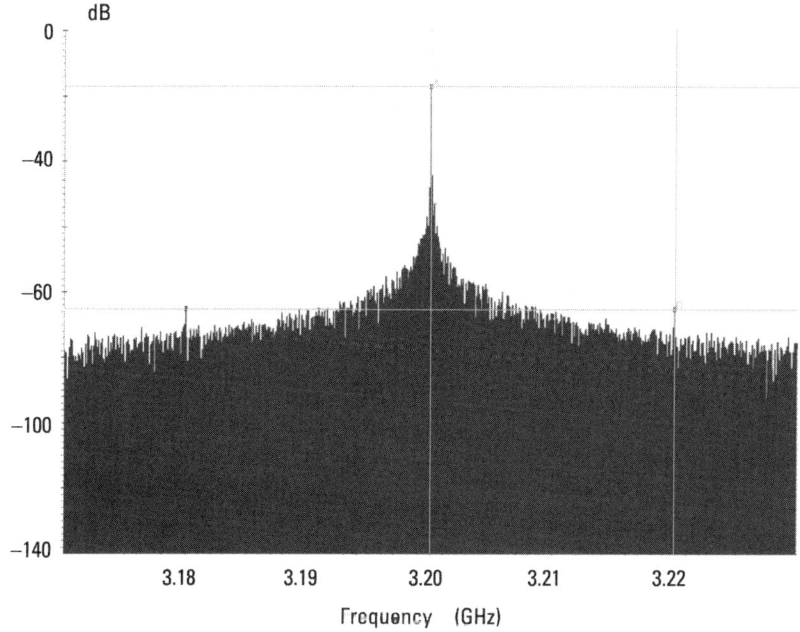

Figure 10.3 DFT of the simulated output signal of the synthesizer.

level of reference spurious emissions of −68.2 dBc. From the output power point of view, the amplitude at the output of the LC-tank is estimated to be 2.08V using the gain of the buffer of the VCO. The phase noise at the output of the synthesizer, the integral of this noise and the lock time of the loop have been obtained using the Simusyn tool. The inputs chosen are contained in Table 10.3, and the variation of the phase noise according to the offset frequency illustrated in Figure 10.4 has been obtained.

Note that the noise reference mask has already been discussed in Section 6.1.3, the phase noise of the VCO in Section 7.2.2.3, the noise of frequency divider in Section 8.4 and the noise L_{1Hz} of the phase detector in Section 9.2.3. The results of their simulations have been chosen for this calculation.

In the graph in Figure 10.4 it can be seen that the phase noise of the synthesizer at 1 MHz of offset practically coincides with the phase noise of the VCO at the same frequency, reaching −119 dBc/Hz. As it has been said previously this is totally reasonable given that at frequencies far from the bandwidth of the loop the VCO constitutes the dominant source of noise.

At the same time the integral of the phase noise reaches a value of −32.6 dBc, which is less than the required value (−32 dBc).

Finally, considering the synthesizer to be variable and assuming a maximum frequency jump of 200 MHz with a tolerance of 1 kHz, an estimation of the

Table 10.3
Input Parameters to Simusyn to Calculate the Phase Noise and the Lock Time of the PLL

Parameter	Value
N	160
R	1
F_{xtal}	20 MHz
$ATTEN$	20 dB
K_{VCO}	170 MHz/V
ϕ_p	55°
L_{1Hz}	−202.9 dBc/Hz
f_c	20 kHz
F_{out}	3.2 GHz
K_ϕ	50 μA
$\lvert f_1 - f_2 \rvert$	200 MHz
tol	1 kHz
L_{Xtal} at 1 kHz	−128.7 dBc/Hz
L_{VCO} at 1 MHz	−119.1 dBc/Hz
$L_{Divisor}$ at 1 kHz	−147.6 dBc/Hz

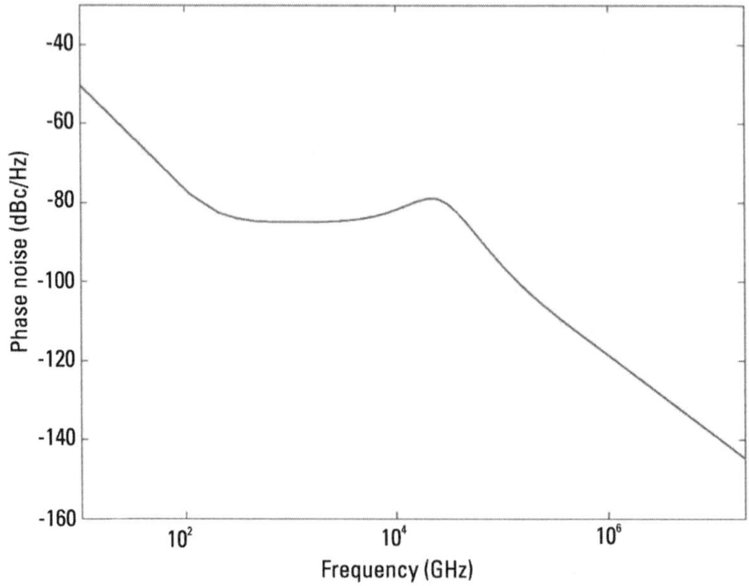

Figure 10.4 Phase noise of the synthesizer at schematic level.

lock time has been obtained using Simusyn. Its result is 0.15 ms which is less than the initial requirement determined by the standard.

In order to have all the results of the synthesizer at schematic level readily available they are contained in Table 10.4.

Once it has been verified that all the results of the synthesizer meet the initial requirements, the design of the layout of this component will be presented in the next section.

10.3 Layout of the Synthesizer

As with the design of the schematic circuit, in order to implement the global layout of the frequency synthesizer the layouts corresponding to each one of its three principal components have been combined. As seen in Section 10.2, the high value of the components of the loop filter make it necessary to be external, which is why it has not been included inside the layout and will be presented as a separate block in the section on experimental results.

At the time of interconnecting the layouts of the three blocks, special care must be taken with the components that work at a higher frequency and with those that present differential inputs or outputs. Therefore, the most critical interconnection is the union between the VCO and the frequency divider. The frequency of output of the VCO of 3.2 GHz together with the need for this interconnection to be differential makes it essential to maintain the maximum symmetry possible between the two tracks of connection. In addition, to minimize the resistance, capacitance and parasitic inductance introduced, the distance between these has been reduced as far as possible using the most external layer (Metal6 in our case) available. Finally, avoiding the crossings between tracks of metal and the excessively close tracks of parallel metal has been attempted in order not to introduce parasitic capacities between them that could degrade the design by effect of crosstalk.

Table 10.4
Results of the Simulation of the Synthesizer at Schematic Level

Parameter	Value
Frequency of output (GHz)	3.2
Phase noise at 1 MHz (dBc/Hz)	–118.85
Integral of the total phase noise on $BW_{channel}$ (dBc)	–32.6
Level of spurious emissions of reference (dBc)	–68.15
Lock time (ms)	0.15
Differential amplitude at the output of the LC-tank (V)	2.08

The rest of the tracks of interconnection, despite not being so critical as the previous ones, need to be treated equally in order to optimize the characteristics of the synthesizer and to achieve a final layout as compact as possible.

Once the layouts of the three blocks have been connected, the following additional components have been laid out: Connection pads, contacts to ground in the substrate around the entire synthesizer, an open ground ring and another supply ring around the circuit and filtering capacitors between V_{DD} and ground. With all these components included, the final layout of the synthesizer can be seen in Figure 10.5, which includes of twelve pads with the following distribution:

- The superior central pad is used to bias the synthesizer except for the buffer of the VCO: V_{tank} in Figure 10.5.
- The inferior side pads constitute the differential outputs of the synthesizer: OUT+ and OUT– in Figure 10.5.
- The superior left is the reference input (X_{tal}), and the inferior left pad the output of the charge pump (CP_{out}).
- The superior right constitutes the voltage control of the VCO (V_{tune}) and the inferior serves to bias the buffer of this oscillator (V_{buffer}).
- The rest of the pads are ground connections: all the GND in Figure 10.5.

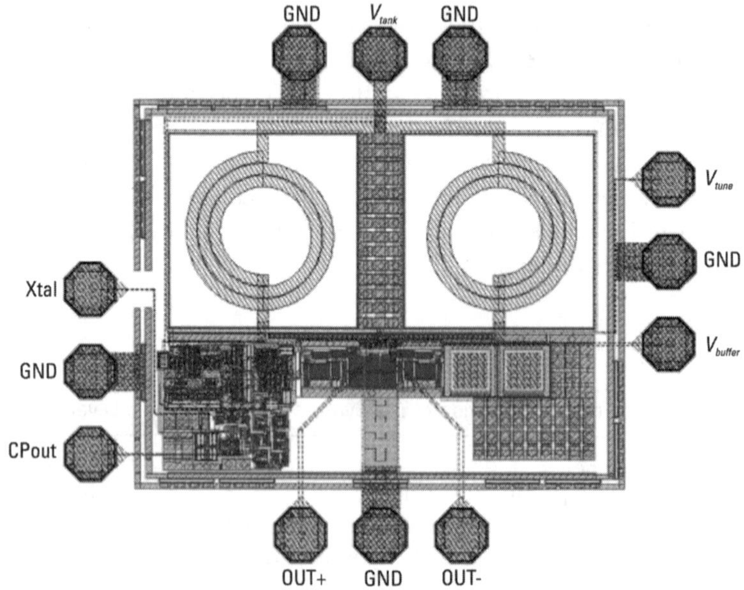

Figure 10.5 Layout of the frequency synthesizer.

Having presented the design of the layout, the results of the postlayout simulation of the component will now be explained followed by the verification of the fulfilment of all the initial requirements contained in Table 10.1. The values of the components of the filter of the loop used in the post layout simulation are those presented in Table 10.2.

As with the schematic circuit design, a transient simulation has been carried out in order to check that the loop locks correctly. From the result of the DFT it can be derived that the output frequency of the locked synthesizer is effectively 3.2 GHz and with a power of 1.69 dBm, which means an output differential amplitude in the LC-tank of 1.81V.

Following the same procedure explained in the schematic design stage (10.2), and using the parameters contained in Table 10.5 in the Simusyn software, the behavior obtained for the phase noise is shown in Figure 10.6.

In this case, the phase noise of the VCO, the frequency divider and noise $L_{1\text{Hz}}$ of the phase detector have been extracted from the postlayout simulation of each one of the corresponding circuits. On the other hand, the noise of the reference crystal is the same as that used in Section 10.2.

In this last figure it can be seen that the value of total phase noise of the synthesizer in postlayout simulation is −118.56 dBc/Hz, less than the initial requirement established (−110 dBc/Hz). The value of the integral of this noise on the bandwidth of the channel (20 MHz) using a bandwidth of the loop of

Table 10.5
Parameters Used to Calculate the Phase Noise and Lock Time of the Frequency Synthesizer of the Postlayout Simulation

Parameter	Value
N	160
R	1
F_{xtal}	20 MHz
$ATTEN$	20 dB
K_{VCO}	170 MHz/V
ϕ_p	55°
$L_{1\text{Hz}}$	−202.3 dBc/Hz
f_c	20 kHz
F_{out}	3.2 GHz
K_ϕ	50 μA
$\lvert f_1 - f_2 \rvert$	200 MHz
tol	1 kHz
L_{Xtal} at 1 kHz	−128.7 dBc/Hz
L_{VCO} at 1 MHz	−118.8 dBc/Hz
$L_{Divisor}$ at 1 kHz	−145 dBc/Hz

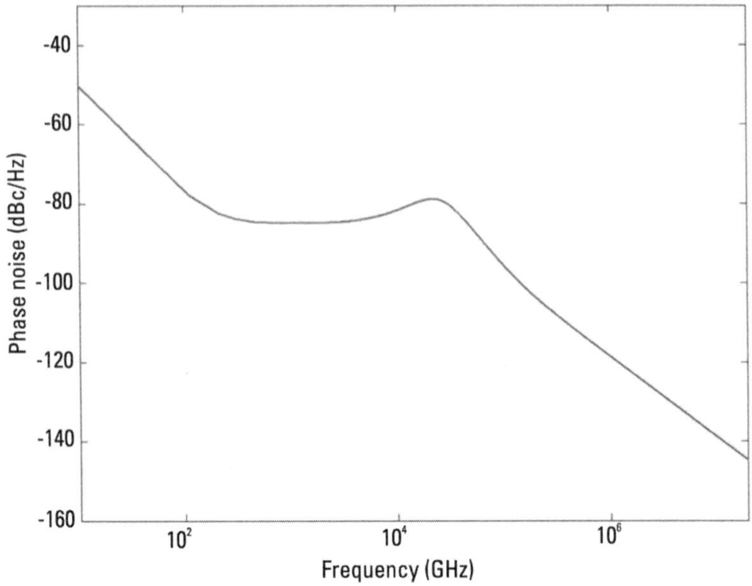

Figure 10.6 Phase noise of the synthesizer at postlayout level.

20 kHz is of the order of −32.3 dBc, which is slightly less than the required level (−32 dBc). Nevertheless, even if the margin from the final simulation to the requirement seems to be close to zero, this figure is appropriate because the main variation of the phase noise curve in the final measurement stage is produced at the highest frequency offsets (mainly influenced by the VCO, as seen in other chapters). This means in the lowest values of PN, and consequently the worsening of the integral of this phase noise on the bandwidth is negligible.

Table 10.6 summarizes the results obtained from the requirements, the schematic simulation and, finally in the third column, the postlayout simulation of the frequency synthesizer. All the parameters fulfill the requirements stated and it is also interesting to note that the variation from the schematic to the postlayout simulations of the circuit is minimal. This result validates, at this stage, the layout realization performed.

Table 10.6
Results of the Postlayout Simulation of the Synthesizer

Parameters	Requirements	Schematic Simulations	Postlayout Simulations
Output frequency (GHz)	3.2	3.2	3.2
Phase noise at 1 MHz (dBc/Hz)	–110	–118.85	–118.56
Integral of the total phase noise over $BW_{channel}$ (dBc)	–32	–32.6	–32.3
Lock time (ms)	1	0.15	0.15
Differential amplitude at the output of the LC-tank (V)	0.5	2.08	1.81
Power consumption (mW)	—	100	100.2

11

PLL Characterization and Results

This chapter shows the characterization process carried out to the fabricated PLL and the individual blocks, and the results obtained for all. Although the text is focused on a particular example, the structure of the chapter follows the shape of a final report for the circuit designed. Regardless of the industrial or academic character of the project we work on, all designers should have the clear idea that the circuit is not finished until the moment the final documentation is written. For this reason, the final section of the chapter is a discussion of the obtained results, where we analyze our design compared to the initial expectations and to other reported circuits.

11.1 VCO

The experimental setup required to characterize the VCO is shown in Figure 11.1. The characteristics of the equipment used for this task is included in Table 11.1, together with the rest of the devices used in the characterization of the complete PLL.

Figure 11.2 shows a microphotograph of the NMOS VCO designed in this work, which occupies an area of $850 \times 1,000\ \mu m^2$. As it is apparent from this graph, one RF probe (output signal) and two DC probes (voltage supply, and VCO voltage control) are required to perform the measurements. The layout view of this microphotograph is shown in Figure 7.15.

The de-embedding of all the elements present in this measurement setup, and in any of the ones shown along this chapter, has been performed using the concepts presented in the application Note 1364-1 from Agilent [1]. All the

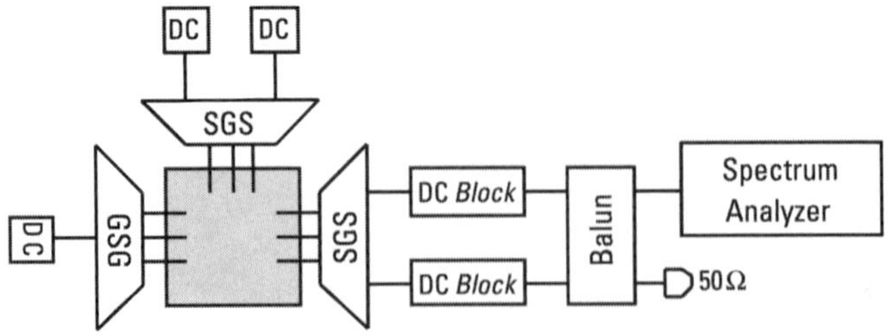

Figure 11.1 VCO characterization diagram.

Table 11.1
Elements Used for the Measurement Carried Out in the Example of This Book

Element	Manufacturer	Characteristics
Probe Station	Cascade Microtech	
SGS probe ACP40 150 μm pitch	Cascade Microtech	Up to 18 GHz
GSG probe ACP40 150 μm pitch	Cascade Microtech	Up to 18 GHz
Optics SZ60	Olympus	
Calibration kits 85052D / 85056A	Agilent	smd and 2.4mm
Spectrum analyzers E4440A/E4407B	Agilent	Up to 26.5 GHz
Signal generator E4438C	Agilent	Up to 6 GHz
Multimeter 2000	Keithley	
Hybrid Coupler (balun) 3A0056	Anaren	2–4 GHz
DC Supply sources E3631A	Hewlett Packard	Double output
Cables Sucoflex 104A	Suhner	0.6m, 18 GHz
Cables 4899	Rosenberger	1.2m, 18 GHz
SMA 90° connectors 23 SMA-50-0-51/199 NE	Suhner	Up to 18 GHz
DC Blocks 7006	Weinschel Corp.	10 kHz–18 GHz
50Ω loads M1406	Weinschel Corp.	Up to 18 GHz
Reference crystal MFO-280F	KSS	20 MHz, ±15 ppm

losses and frequency dependant effects have to be taken away from the final measurement.

Figure 11.3 presents the tuning range of the VCO when the voltage control V_{tune} is varied from 0.5V to 2.8V. In addition, this range is compared with the results obtained from the postlayout simulation.

The clearest difference of the curves illustrated in this figure is the displacement of the measured range towards frequencies slightly lower than those estimated by the electrical circuit simulation tool. This is mainly due to the fact that the parasitic capacitances and inductances introduced in the oscillator are slightly greater than those estimated in the design. However, this is not

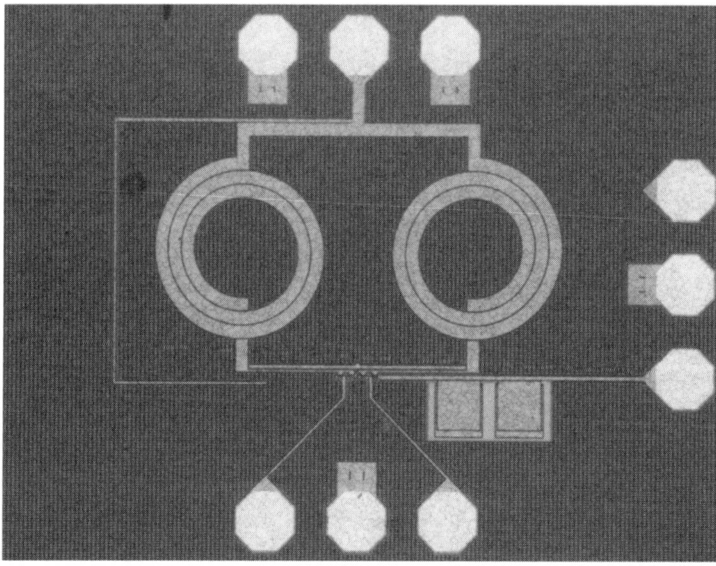

Figure 11.2 Microphotograph of the NMOS VCO.

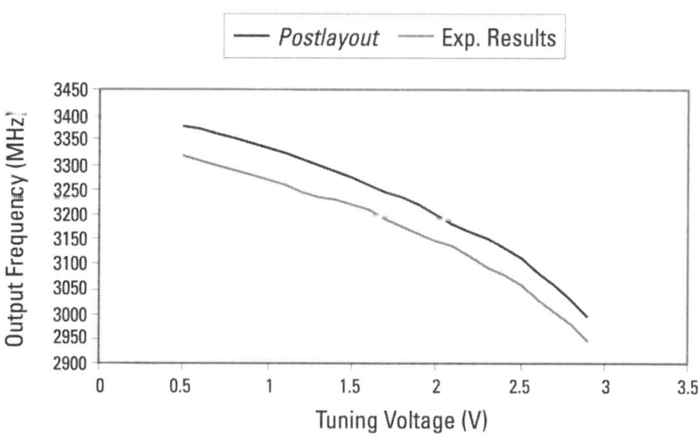

Figure 11.3 Measured and simulated tuning range of the VCO.

problematic. In fact, even with this displacement it has been possible to reach the desired frequency of oscillation of 3.2 GHz at a voltage control of approximately 1.64V, which is practically the central point between the extremes 0.5V and 2.8V. It has already been described that a security/safety margin was adopted in the postlayout simulation to compensate for this effect. Furthermore, by reducing the VCO gain the amplification of the noise voltage present in the VCO line of control is attenuated. This gain, which presents a value in the postlayout simulation in the order of 170 MHz/V, becomes approximately 144 MHz/V, as can be seen in Figure 11.3. It is important to point out that for this first measurement a value of 3.3V has been used for the supply voltages as much for the tank as for the output stage.

The output power measured at 3.2 GHz and 3.3V of supply voltage is approximately 1 dBm, which is slightly less than that obtained in the postlayout simulation. From this result and using the known values of gain of the output stage and of the coupling factor of the balun, it is possible to estimate the differential amplitude at the output of the tank at approximately 1.7V. It is interesting to note at this point that a balun is needed in the measurement setup to convert the differential output signal to a single-ended signal suitable for spectrum analyzer measurements. Evidently, the balun has to be selected taking into consideration the frequency, power and isolation data of this specific measurement. In this case, to work in the 3.2-GHz band, the one from 2 to 4 GHz is used, with an available input power greater than 3 dBm (in this case its maximum is 27 dBm or 0.5W) and with the greater isolation possible among the ports (higher than 20 dB).

Regarding the phase noise measurements, the noise coming from the DC sources and other interferers is coupled to the oscillator through the supply and control tracks of the VCO causing considerable instability in the frequency. This makes it difficult to obtain a reliable measurement of phase noise. Due to this, it is necessary to measure the phase noise of the VCO inside the PLL in locked loop condition. Figure 11.4 shows this characterization at frequencies above 1 MHz, where the VCO is dominant. According to this figure the oscillator possesses a phase noise of −118 dBc/Hz at 1 MHz of *offset*, which as it has yet been noted in Chapter 10, is 8 dB better than called for in the specifications as listed in Table 7.1.

Finally, Table 11.2 contains the summary of the specified values, simulated in postlayout and measurements of the oscillator. It can be observed that they all fulfil the set of specifications required.

The difference in output power is mainly due to the effect of series resistances not considered by the simulation tool. Finally, with reference to the phase noise it must be pointed out that the simulator only takes into account the zone of the 20 dB/decade gradient, and does not simulate the effect of the flicker noise or the conversion of the thermal noise. Furthermore, the simulation tool does not take into account the noise introduced through the supply sources either.

11.2 Frequency Divider

This second section focuses on presenting the most important experimental results from the frequency divider followed by the verification of the fulfillment

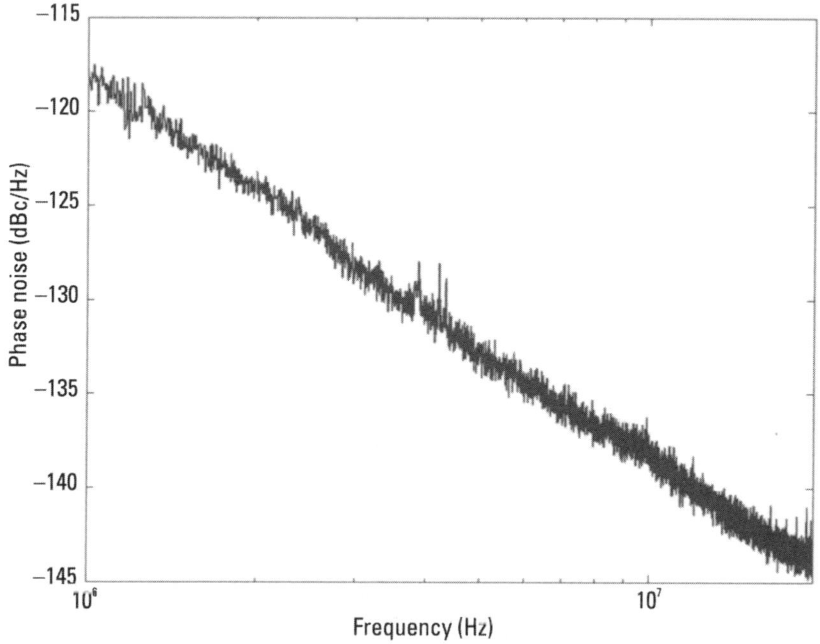

Figure 11.4 Phase noise of the NMOS VCO.

Table 11.2
Comparison Between Specifications, Simulations, and Measurements of the VCO

	Specified	Simulated	Measured
Frequency of central oscillation (MHz)	3,200	3,250	3,200
Tuning range (MHz)	322	360	351
Phase noise at 1 MHz (dBc/Hz)	−110	−118.6	−118
Output power (dBm)	—	2	1
Power consumption (mW)	—	83	82.5
Differential amplitude (V)	>0.5	1.93	1.7

of its initial specifications. The measurement diagram used to characterize the frequency divider is shown in Figure 11.5, and is linked to the elements presented in Table 11.1.

Figure 11.6 presents a microphotograph of the frequency divider implemented, which occupies an area of 661 × 733 μm^2 including the *pads* of connection. The layout view of this microphotograph is shown in Figure 8.14.

The results of characterization that are presented in this section are the output frequency, the value of the module of division, the maximum frequency

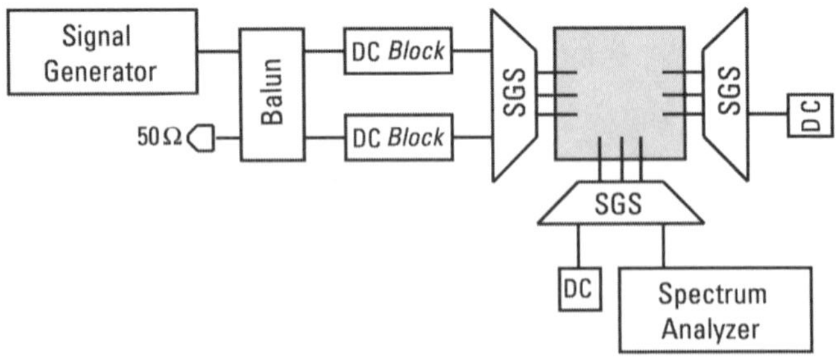

Figure 11.5 Measurement diagram of the frequency divider.

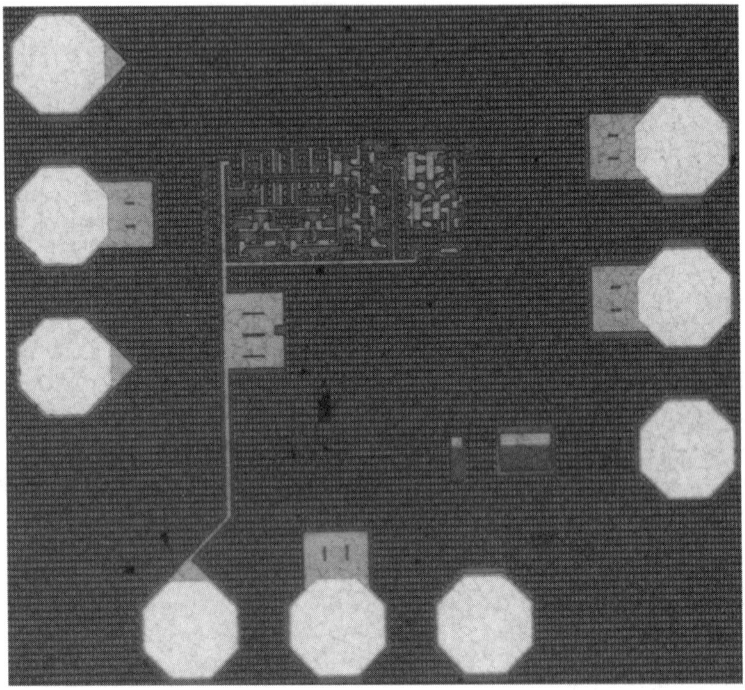

Figure 11.6 Microphotograph of the frequency divider.

of operation and the minimum working amplitude in each one of its two differential inputs.

First, the process followed to obtain the value of its module of division is demonstrated. For this purpose a 3.2-GHz sinusoidal signal has been introduced in the input of the divider and the output has been measured using a spectrum analyzer. The result is shown in Figure 11.7.

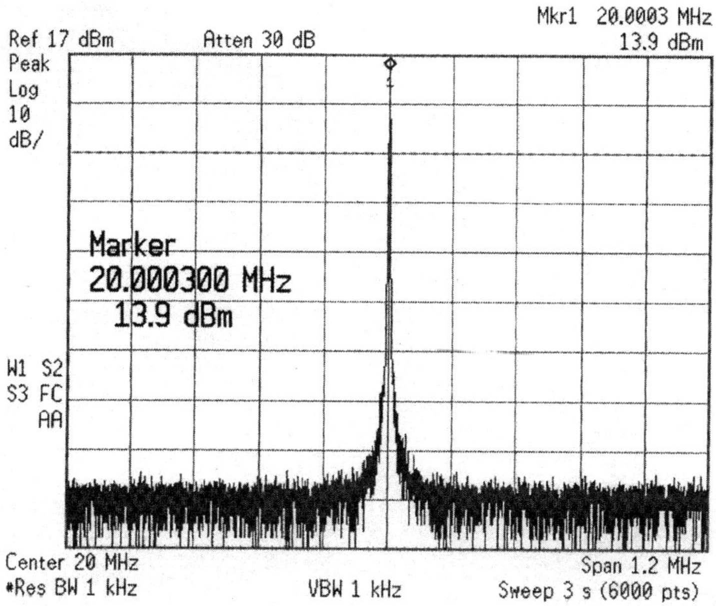

Figure 11.7 Output of the frequency divider measured with the spectrum analyzer.

As the frequency of the output signal of the divider is 20 MHz for an input of 3.2 GHz, it can be asserted that the value of the rate of division is 160, fulfilling the value specified.

Despite not forming part of the specifications of the component, it is interesting to present the experimental results referring to the maximum working frequency of the divider for an input power of 0 dBm, and the minimum input signal at 3.2 GHz for which the divider operates correctly.

The maximum working frequency for an input power of 0 dBm is approximately 3.66 GHz. This result can be verified by consulting Figure 11.8, in which the output signal of the divider for the maximum input frequency mentioned earlier is shown. Considering that the rate of division is 160, it can be estimated that the value of the input frequency to the divider is approximately 3.66 GHz.

At the same time, the minimum input power for which the divider functions correctly is −16.1 dBm, which means a power of −19.1 dBm for each one of the two differential inputs. From this minimum power and using the electrical circuit simulator it has been possible to calculate the amplitude of each one of the differential inputs of the divider corresponding to this value of power, resulting in 66 mV. This amplitude depends logically on the input impedance of this component.

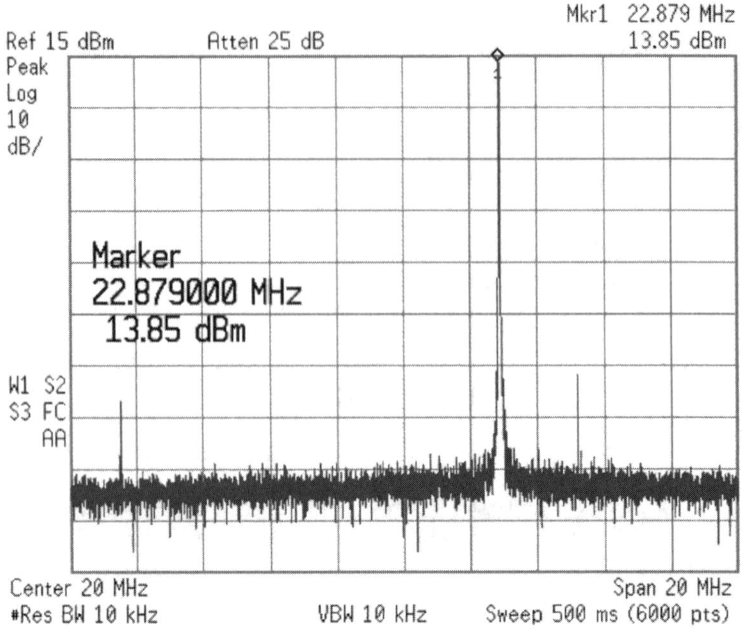

Figure 11.8 Output of the divider for the maximum input frequency.

To conclude this section, Table 11.3 contains the summary of the values of the frequency divider, simulated in postlayout and measured.

It is important to point out that the phase noise obtained from the postlayout simulation is negligible compared to the rest of the sources of noise inside the loop estimated in the initial simulation of the system.

11.3 Complete PLL

The diagram used to obtain the experimental results of the frequency synthesizer is shown in Figure 11.9, where the LPF block represents the external low-pass

Table 11.3
Comparison Between Specifications, Simulations, and Measurements of the Divider

	Simulated	Measured
Measured value of the division modulus	160	160
Phase noise at 1 kHz (dBc/Hz)	–145	—
Power consumption (mW)	19.1	19.8
Maximum frequency (GHz)	3.8	3.66
Minimum input amplitude (mV)	51	66

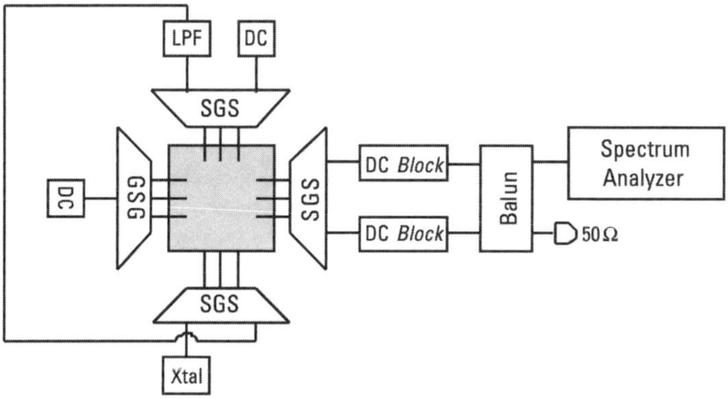

Figure 11.9 Characterization diagram of the frequency synthesizer.

filter of the loop. This filter has been implemented in an auxiliary PCB, and the values that have been chosen for their components are the same as those calculated with Simusyn. The other blocks here shown have been listed in Table 11.1.

Figure 11.10 shows a photograph of the setup arranged in order to obtain these experimental results.

In addition to this setup, Figure 11.11 shows a more detailed photograph of the RF probes used.

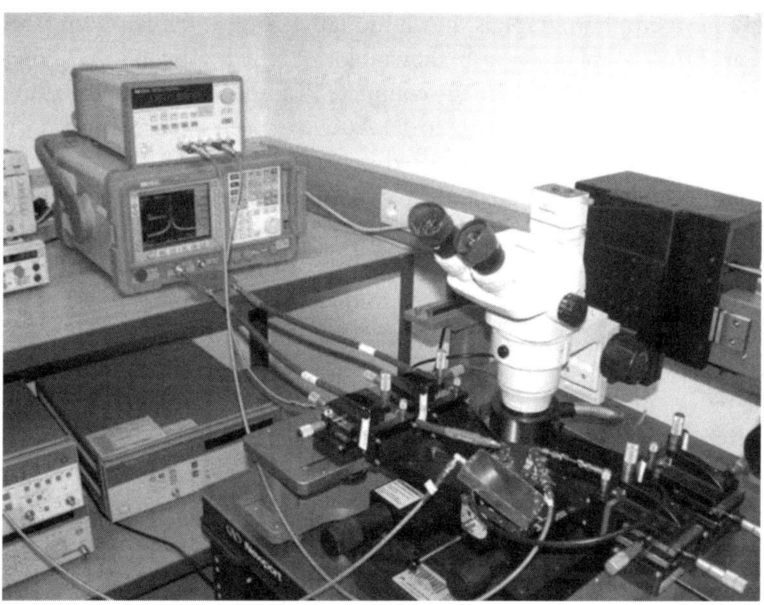

Figure 11.10 Photograph of the measurement setup.

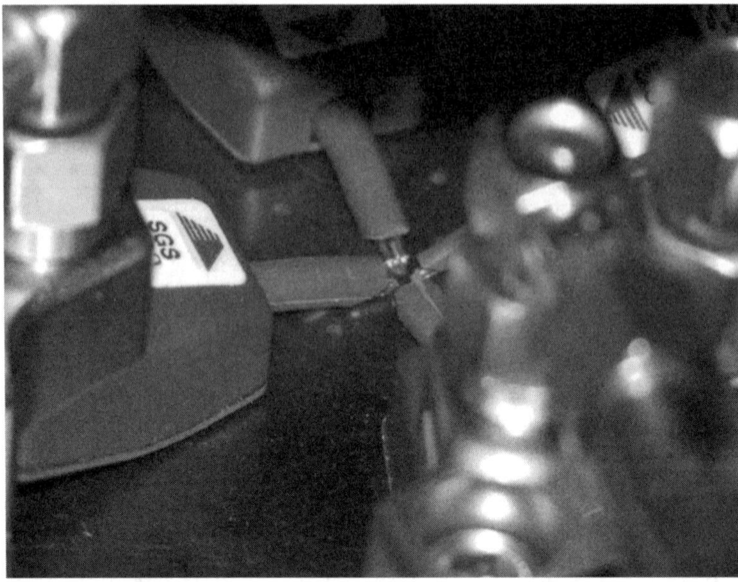

Figure 11.11 Photograph of the RF probes.

Before presenting the experimental results, a microphotograph of the frequency synthesizer implemented in the work presented in this book, which possesses an area of 1,000 × 1,200 μm^2, is shown in Figure 11.12. The layout view of this microphotograph is shown in Figure 10.5, and it has been fabricated at the same time as the previously shown blocks, which validates the simulations presented in Chapter 10, where the complete PLL was formed by the simulation results of the blocks that composed it. As it has been exposed previously, the external loop filter has been connected between CP_{OUT} and V_{tune} terminals.

The results demonstrated in this section are those corresponding to the output frequency, phase noise, its integral over the channel bandwidth, the level of spurious emissions, and the output power of the PLL, which has facilitated the estimation of the differential amplitude in the LC-tank.

First, Figure 11.13 shows the screen of the spectrum analyzer corresponding to the locked PLL at the frequency of operation of 3.2 GHz. In addition, the output power of the synthesizer is 0 dBm approximately, which corresponds to a differential amplitude in the LC-tank of 1.5V.

But the most important result shown in Figure 11.13 is the level of the reference spurious emissions, which reaches a value of −64.3 dBc.

As described in Section 2.2.2, the level of these spurious emissions depends notably on the constant *BasePulseSpur*. From the experimental result of the reference spurs and using the Simusyn tool the value of this constant has been estimated at −283 dBc.

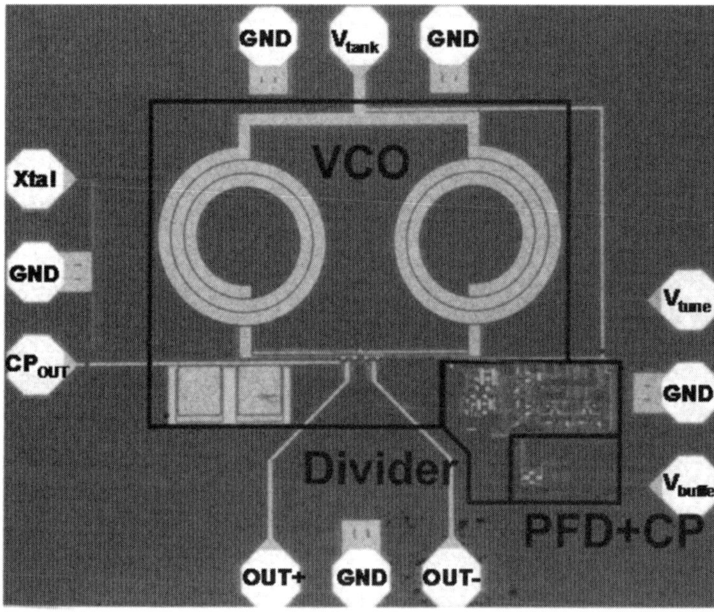

Figure 11.12 Microphotograph of the frequency synthesizer.

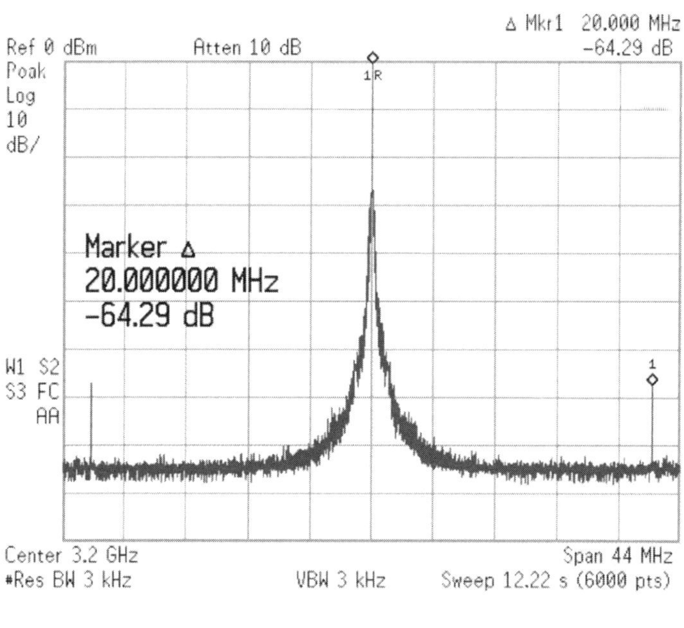

Figure 11.13 Capture of the screen of the spectrum analyzer with the PLL locked.

Figure 11.14 illustrates the phase noise of the frequency synthesizer together with the phase noises corresponding to the initial and final simulation of the system. It is important to point out that this measured phase noise reaches a value of −118 dBc/Hz at 1 MHz. This value differs from that corresponding to the initial simulation of the system for the same frequency (−110 dBc/Hz), due to the conservative assumptions taken during the circuit design. In this graph we also show the comparison among three situations:

- System simulation with Simusyn, when the input data are based on predictions;
- System simulation with Simusyn, with input data corrected with building block characterization;
- Characterization results.

From the graphs obtained we can conclude that Simusyn offers an acceptable first prediction of phase noise based just on simulation data. However, when these inputs are adjusted with real results of the building blocks of the PLL, the prediction improves. This can be very useful when trying to combine already developed building blocks for new applications.

Finally, the value of the integral of phase noise measured over the channel bandwidth (20 MHz) is approximately –32.2 dBc.

In order to be able to make a better comparison of all the parameters of the frequency synthesizer the experimental results obtained together with

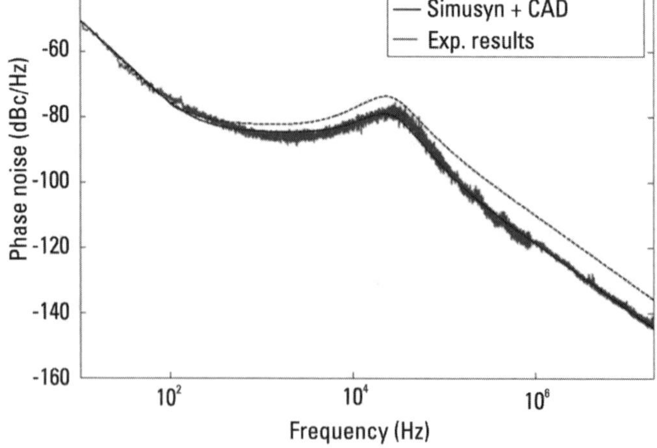

Figure 11.14 Measured phase noise of the frequency synthesizer.

those of the postlayout simulation and the initial requirements are contained in Table 11.4.

11.4 Result Discussion

Once the characterization of the circuit has been carried out, it is time to review the whole process and discuss the obtained results. This should be the final stage in the design process, that is, the work is finished when we realize the strong and weak points of our work, and we extract conclusions that will be applied in our next challenge. A good discussion should have at least two points:

- Comparison of the results with previously reported works;
- Analysis of the validity of the work for the targeted application.

Regarding the comparison with other results, Table 11.5 presents different PLLs developed for the IEEE 802.11a standard in CMOS technology. These characteristics are, from left to right, the output frequency, the fabrication technology, the phase noise, the level of reference spurs, the voltage supply, power consumption, the output power of the PLL, and the integral of phase noise over a bandwidth of 10 MHz.

It is worth pointing out that the phase noise of the different frequency synthesizers in this table has been normalized at the frequency of operation of the PLL implemented in the example presented in this book (3.2 GHz) according to (11.1).

$$L_{norm}\{\omega\} = L\{\omega\} + 20\log\left(\frac{\omega_0}{\omega_0^*}\right) \tag{11.1}$$

Table 11.4
Comparison Between Specifications, Simulations, and Measurements of the Synthesizer

	Specified	Simulated	Measured
Output frequency (GHz)	3.2	3.2	3.2
Phase noise at 1 MHz (dBc/Hz)	–110	–118.56	–118
Integral of the phase noise on BW_{canal} (dBc)	–32	–32.3	–32.2
Reference spurs (dBc)	–59	–68.15	–64.29
Amplitude in the LC-tank (V)	>0.5	1.81	1.5
Consumption (mW)	—	100.2	99

Table 11.5
Reported PLLs for the IEEE 802.11a Standard

Reference	Freq. (GHz)	Tecn. CMOS	Phase Noise at 1 MHz (dBc/Hz)	*Spurs* (dBc)	V_{dd} (V)	Cons. (mW)	Pot. (dBm)	∫P. N. (dBc)
[2]	5.5	0.13 μm	–116.7	—	1.2/2.5	36	—	–36
[3]	4.3	0.25 μm	—	–58 at 4 MHz	2.5	117.5	—	—
[4]	4.9	0.25 μm	–104.7	–54 at 22 MHz	1.5/2	25	—	—
[5]	3.5	0.18 μm	–120.7	–66 at 13 MHz	1.8	—	—	–31.7
[6]	5.2	0.18 μm	–114.2	—	1.8	—	—	—
[7]	5.2	0.18 μm	–119.2	–65 at 10 MHz	1.8	—	–7.14	–32
[8]	5.5	0.18 μm	–115.7	–80 at 11 MHz	1	27.5	–5.5	—
[9]	4	0.25 μm	–114	—	2.5	180	—	–33
[10]	5.2	0.18 μm	–116.2	—	1.8	—	—	–34
[11, and this book]	3.2	0.18 μm	–118	–64 at 20 MHz	3.3	100.2	0	–33.73

where $L_{norm}\{\omega\}$ is the normalized phase noise, ω_0 is the frequency at which the comparison to be carried out is desired, and ω_0^* is the real fundamental frequency of the oscillator.

After the analysis of the information contained in Table 11.5, it can be noted that the frequency synthesizer implemented in this research project presents a phase noise at 1 MHz of *offset* approximately 3 dB less than the average (–115.1 dBc/Hz). At the same time, the integral of this noise on a bandwidth of 10 MHz is also slightly less than the average (–33.34 dBc).

The comparison of the level of spurious emissions presents the difficulty that the reference crystal at which they appear presents substantial variations between the different implementations. This depends logically on the configuration chosen to implement each one of them. Despite all this, the synthesizer implemented in the present example has a level of reference spurs close to the average (–64.6 dBc).

In reference to the output power of the PLL, there is not much information available. Compared to the two presented in Table 11.5, the synthesizer implemented in this book possesses a much greater power of output.

The power consumption is slightly higher than the average found in the bibliographical references presented in Table 11.5 (77.2 mW). It must be noted that this circuit has not been optimized for low power consumption, and therefore, there is room for improvement in this point, as it has been discussed in the design of the different building blocks. For example, the straighter way to decrease this power consumption in the global example shown in this book would be to reduce the VCO power consumption which has proved to be at least 70% of the overall power consumption. The phase noise simulation and measurement has been 8 dB better than the requirement, so there is room for much decrease in the active circuit consumption of the VCO.

To end this discussion, in Table 11.4 it can be seen that all the experimental results meet the initial requirements established in Chapter 6. Therefore, it can be asserted that the synthesizer implemented in this work is valid for the 5-GHz band WLAN IEEE 802.11a receiver proposed.

References

[1] Agilent Technologies, "De-Embedding and Embedding S-Parameters Networks Using a Vector Network Analyzer," Application Note 1364-1, October 2004.

[2] Valla, M., et al., "A 72-mW CMOS 802.11a Direct Conversion Front-End with 3.5-dB NF and 200-KHz 1/f Noise Corner," *IEEE Journal of Solid-State Circuits,* Vol. 40, No. 4, April 2005, pp. 970–977.

[3] Hernández, J., et al., "Analysis of Architectures for 1.8 GHz CMOS LC-Tank Voltage-Controlled Oscillators," *Proceedings of DCIS'99,* Palma de Mallorca, November 1999, pp. 139–142.

[4] Rategh, H., H. Samavati, and T. Lee, "A CMOS Frequency Synthesizer with an Injection-Locked Frequency Divider for a 5-GHz Wireless LAN Receiver," *IEEE Journal of Solid-State Circuits*, Vol. 35, No. 5, May 2000, pp. 780–787.

[5] Zhang, P., et al., "A 5-GHz Direct-Conversion CMOS Transceiver," *IEEE Journal of Solid-State Circuits*, Vol. 38, No. 12, December 2003, pp. 2232–2237.

[6] Behzad, A. R., et al., "A 5-GHz Direct-Conversion CMOS Transceiver Utilizing Automatic Frequency Control for the IEEE 802.11a Wireless LAN Standard," *IEEE Journal of Solid-State Circuits*, Vol. 38, No. 12, December 2003, pp. 2209–2220.

[7] Vassiliou, I., et al., "A Single-Chip Digitally Calibrated 5.15-5.825-GHz 0.18-μm CMOS Transceiver for 802.11a Wireless LAN," *IEEE Journal of Solid-State Circuits*, Vol. 38, No. 12, December 2003, pp. 2221–2229.

[8] Leung, G. C. T., and H. C. Luong, "A 1-V 5.2-GHz CMOS Synthesizer for WLAN Applications," *IEEE Journal of Solid-State Circuits*, Vol. 39, No. 11, November 2004, pp. 1873–1882.

[9] Zargari, M., et al., "A 5-GHz CMOS Transceiver for IEEE 802.11a Wireless LAN Systems," *IEEE Journal of Solid-State Circuits*, Vol. 37, No. 12, December 2002, pp. 1688–1694.

[10] Ahola, R., et al., "A Single-Chip CMOS Transceiver for 802.11a/b/g Wireless LANs," *IEEE Journal of Solid-State Circuits*, Vol. 39, No. 12, December 2004, pp. 2250–2258.

[11] Quemada, C., et al., "A CMOS Frequency Synthesizer with Self-Biasing Current Source for 5 GHz Wireless-LAN Receiver," *Microwave Journal*, February 2007.

About the Authors

Carlos Quemada received an M.Sc. in 2000 in telecommunication engineering from the Public University of Navarra (UPNA) and a Ph.D. in 2006 from the University of Navarra's engineering school, collaborating as an assistant professor. After completing his M.Sc. degree, he joined CEIT in 2001 as a researcher on the design and fabrication of high frequency analog integrated phase-locked loops (PLLs) for communication WLAN front ends. Since March 2008, he has worked for the research center IKERLAN inside the communications area. Currently, his research area is focused on the design and implementation of embedded systems for digital communications.

Guillermo Bistué received an M.Sc. and a Ph.D. from the University of Navarra's engineering school in 1994 and 1997, respectively. Having completed his studies, he joined CEIT in 1997 as a researcher and presently leads the Communications Group at the prestigious research center. Dr. Bistué's current research field is focused on the design and fabrication of analog integrated circuits for communication front ends. In the course of his career he has taken part in several industrial and basic research projects, dealing with wireless standards communications like WLAN, DVB-H, GALILEO, or GPS. He is also a lecturer on basic electronics at TECNUN (Engineering School of University of Navarra).

Iñigo Adin received an M.Sc. and a Ph.D. from the University of Navarra's engineering school in 2003 and 2007, respectively. His Ph.D. research was entitled "RF CMOS IC Design applied to Multistandard Wireless Applications for the 5-GHz U-NII Band." He has also worked on analog IC design and fabrication for applications as ESD protection for low power front-ends. Dr.

Adin is currently engaged in the design of advanced communications systems for European high-speed trains; he is also a lecturer on electronic circuit design at TECNUN (Engineering School of University of Navarra).

Index

Recent Titles in the Artech House Microwave Library

Active Filters for Integrated-Circuit Applications, Fred H. Irons

Advanced Techniques in RF Power Amplifier Design, Steve C. Cripps

Automated Smith Chart, Version 4.0: Software and User's Manual, Leonard M. Schwab

Behavioral Modeling of Nonlinear RF and Microwave Devices, Thomas R. Turlington

Broadband Microwave Amplifiers, Bal S. Virdee, Avtar S. Virdee, and Ben Y. Banyamin

Computer-Aided Analysis of Nonlinear Microwave Circuits, Paulo J. C. Rodrigues

Designing Bipolar Transistor Radio Frequency Integrated Circuits, Allen A. Sweet

Design of FET Frequency Multipliers and Harmonic Oscillators, Edmar Camargo

Design of Linear RF Outphasing Power Amplifiers, Xuejun Zhang, Lawrence E. Larson, and Peter M. Asbeck

Design Methodology for RF CMOS Phase Locked Loops, Carlos Quemada, Guillermo Bistué, and Iñigo Adin

Design of RF and Microwave Amplifiers and Oscillators, Pieter L. D. Abrie

Digital Filter Design Solutions, Jolyon M. De Freitas

Distortion in RF Power Amplifiers, Joel Vuolevi and Timo Rahkonen

EMPLAN: Electromagnetic Analysis of Printed Structures in Planarly Layered Media, Software and User's Manual, Noyan Kinayman and M. I. Aksun

Essentials of RF and Microwave Grounding, Eric Holzman

FAST: Fast Amplifier Synthesis Tool—Software and User's Guide, Dale D. Henkes

Feedforward Linear Power Amplifiers, Nick Pothecary

Foundations of Oscillator Circuit Design, Guillermo Gonzalez

Fundamentals of Nonlinear Behavioral Modeling for RF and Microwave Design, John Wood and David E. Root, editors

Generalized Filter Design by Computer Optimization, Djuradj Budimir

High-Linearity RF Amplifier Design, Peter B. Kenington

High-Speed Circuit Board Signal Integrity, Stephen C. Thierauf

Intermodulation Distortion in Microwave and Wireless Circuits, José Carlos Pedro and Nuno Borges Carvalho

Introduction to Modeling HBTs, Matthias Rudolph

Lumped Elements for RF and Microwave Circuits, Inder Bahl

Lumped Element Quadrature Hybrids, David Andrews

Microwave Circuit Modeling Using Electromagnetic Field Simulation, Daniel G. Swanson, Jr. and Wolfgang J. R. Hoefer

Microwave Component Mechanics, Harri Eskelinen and Pekka Eskelinen

Microwave Differential Circuit Design Using Mixed-Mode S-Parameters, William R. Eisenstadt, Robert Stengel, and Bruce M. Thompson

Microwave Engineers' Handbook, Two Volumes, Theodore Saad, editor

Microwave Filters, Impedance-Matching Networks, and Coupling Structures, George L. Matthaei, Leo Young, and E.M.T. Jones

Microwave Materials and Fabrication Techniques, Second Edition, Thomas S. Laverghetta

Microwave Mixers, Second Edition, Stephen A. Maas

Microwave Radio Transmission Design Guide, Trevor Manning

Microwaves and Wireless Simplified, Third Edition, Thomas S. Laverghetta

Modern Microwave Circuits, Noyan Kinayman and M. I. Aksun

Modern Microwave Measurements and Techniques, Second Edition, Thomas S. Laverghetta

Neural Networks for RF and Microwave Design, Q. J. Zhang and K. C. Gupta

Noise in Linear and Nonlinear Circuits, Stephen A. Maas

Nonlinear Microwave and RF Circuits, Second Edition, Stephen A. Maas

QMATCH: Lumped-Element Impedance Matching, Software and User's Guide, Pieter L. D. Abrie

Practical Analog and Digital Filter Design, Les Thede

Practical Microstrip Design and Applications, Günter Kompa

Practical RF Circuit Design for Modern Wireless Systems, Volume I: Passive Circuits and Systems, Les Besser and Rowan Gilmore

Practical RF Circuit Design for Modern Wireless Systems, Volume II: Active Circuits and Systems, Rowan Gilmore and Les Besser

Production Testing of RF and System-on-a-Chip Devices for Wireless Communications, Keith B. Schaub and Joe Kelly

Radio Frequency Integrated Circuit Design, John Rogers and Calvin Plett

RF Design Guide: Systems, Circuits, and Equations, Peter Vizmuller

RF Measurements of Die and Packages, Scott A. Wartenberg

The RF and Microwave Circuit Design Handbook, Stephen A. Maas

RF and Microwave Coupled-Line Circuits, Rajesh Mongia, Inder Bahl, and Prakash Bhartia

RF and Microwave Oscillator Design, Michal Odyniec, editor

RF Power Amplifiers for Wireless Communications, Second Edition, Steve C. Cripps

RF Systems, Components, and Circuits Handbook, Ferril A. Losee

Stability Analysis of Nonlinear Microwave Circuits, Almudena Suárez and Raymond Quéré

System-in-Package RF Design and Applications, Michael P. Gaynor

TRAVIS 2.0: Transmission Line Visualization Software and User's Guide, Version 2.0, Robert G. Kaires and Barton T. Hickman

Understanding Microwave Heating Cavities, Tse V. Chow Ting Chan and Howard C. Reader

For further information on these and other Artech House titles, including previously considered out-of-print books now available through our In-Print-Forever® (IPF®) program, contact:

Artech House Publishers
685 Canton Street
Norwood, MA 02062
Phone: 781-769-9750
Fax: 781-769-6334
e-mail: artech@artechhouse.com

Artech House Books
46 Gillingham Street
London SW1V 1AH UK
Phone: +44 (0)20 7596 8750
Fax: +44 (0)20 7630 0166
e-mail: artech-uk@artechhouse.com

Find us on the World Wide Web at: www.artechhouse.com